The Hammett Equation

Cambridge Chemistry Texts

The Hammett Equation

C. D. JOHNSON
Lecturer in Organic Chemistry, University of East Anglia

CAMBRIDGE
At the University Press 1973

Published by the Syndics of the Cambridge University Press
Bentley House, 200 Euston Road, London NW1 2DB
American Branch: 32 East 57th Street, New York, N.Y. 10022

© Cambridge University Press 1973

Library of Congress Catalogue Card Number: 72-93140

ISBN: 0 521 20138 1

Printed in Great Britain by
William Clowes & Sons Limited, London, Colchester and Beccles

Contents

Preface

Since its conception over thirty years ago, the Hammett equation together with subsequent modifications, all of which owed their motivation to Hammett's original idea, has provided the main basis for quantitative structure reactivity relationships in organic chemistry. The uses of the equation in the great number of research publications employing it can be broadly divided into two categories.

The first is the elucidation of intramolecular interactions of electronic or steric type, typically the manner in which substituents exert their influence on rates of reactions or positions of equilibria.

The second is the investigation of reaction mechanisms, where it is frequently and to best advantage used in conjunction with other physicochemical techniques, all combining to form a consistent and consequently convincing rationale of the reaction pattern in question.

This book is an attempt to explain these two approaches (and to emphasise the importance of the second) at a level suitable for third year chemistry undergraduates or first year graduate students. Such students, although familar with fundamental qualitative organic chemistry, have often had little or no experience of quantitative assessments of the reactivity of organic molecules. For this reason, exhaustive lists of parameters accompanied by rigorous statistical analysis, possibly of some value to the initiated, but overwhelmingly confusing and indigestible to the beginner, have been avoided. Attention is restricted to detailed consideration of a few familiar reactions and substituents.

However, the simple postulates thus developed can be readily applied to more complicated cases; for this reason, each chapter concludes with a series of problems, taken mostly from recent research papers, and designed to help the student learn to apply the general points included in the text to specific examples of current interest. Perhaps the student will be tempted to try at least some of these for himself. It is impossible to appreciate the concepts of any aspect of physical organic chemistry by lecture attendance and textbook reading alone. Real understanding can only come from supplementation by exercises in which the student calculates and interprets data by himself or possibly in a small discussion group. Through such a process he becomes familiar with the practical meaning to be attached to quantities denoted symbolically in general equations, and thus forms a realistic assessment of the extent of accuracy and degree of validity which he is prepared to accord to such equations.

Consideration of the thermodynamic basis of the equation is delayed until the final chapter. In an elementary treatment, students are prepared to accept the ideas of inductive, resonance, and steric effects without too much questioning. It is only later, when some time has been spent dealing with quantitative approaches such as the Hammett equation or molecular orbital calculations, that they really begin to enquire more critically into the relevance of these concepts, and how and why such theoretical postulates relate to experimentally determined data such as free energy, enthalpy, and entropy changes.

It is impossible in any description of the Hammett equation to avoid controversial material. All aspects are currently under scrutiny; there are many different opinions and conflicting views. In several instances here, points of uncertainty are indicated, but perhaps a degree of simplification is inevitable, for a student must learn what the ideas are before he can effectively appraise them.

Use of the Hammett equation has frequently encountered censure on the grounds that it is empirical, inaccurate, and that the wide diversity of σ value types is confused and ridiculous. There is undoubtedly some truth in this. Certainly one finds quite frequently a measure of significance placed on small order terms which puts credulity under severe strain. However, only the simplest of molecules *in vacuo* can be treated with absolute accuracy; the complete understanding of the complicated systems of organic reactions in solution is a far distant goal, and at the present time theoretical organic chemistry in general is necessarily semi-empirical and approximate. Nevertheless, in terms of the role which it has played in the broad elucidation of electronic effects and reaction mechanisms, the Hammett equation has been of unparalleled utility, to the recognition of which, it is hoped, this volume will in a small measure contribute.

I am most grateful to Dr K. Schofield for all his hard work during preparation of the manuscript for the press. I would also like to thank Dr R. A. Y. Jones for reading the complete text and making many helpful comments, Professors J. C. Barborak, L. Hepler and G. Marino for discussion, and Mary Ellam and Elly Browne for typing the final draft. Finally, my thanks are due to my wife, for her patience, help, and encouragement.

C.D.J.

University of Alberta, Edmonton 1971
University of East Anglia 1972

1 The Hammett σρ relationship

1.1. Introduction. The student of organic chemistry is introduced early to the theory of two distinct and apparently mutually independent modes for transmission of electronic effects, namely induction and resonance (mesomerism). Even if their exact origin and manner of propagation remain uncertain in anything but very simple terms, their terminology at least becomes extremely familiar: moreover, such concepts, despite their limitations, are seen to be the foundation-stones of theoretical interpretation of the mechanisms of organic reactions. Proximity effects and steric interactions can confuse the issue, but although these are exceedingly intricate in origin, the possibility of their intervention is apparent from consideration of the stereochemistry of the molecules involved. In their absence, the effect of a given substituent on reaction rates or equilibrium constants, in terms of its inductive and resonance interaction, is effectively consistent, *qualitatively* at least, through the whole spectrum of reaction types.

The basic aim of the Hammett equation is to evaluate the degree of this consistency in *quantitative* terms. This involves elucidating the contribution of inductive and resonance effects to free energy changes on going from one side to the other of an equilibrium or from the ground state to the transition state of a reaction. Such free energy changes will be proportional to the logarithms of rate or equilibrium constants.

1.2. The Hammett equation. The effect of substituents in the benzene nucleus on the acidity of benzoic acid expressed in terms of pK_a, and the explanation of such effects in terms of induction and resonance, forms a specific example of the ideas generally expressed above. It is also particularly convenient in that, for *m*- and *p*-substituted benzoic acids, electronic factors are very unlikely to be complicated by steric interactions between substituent and reaction site. A quantitative measure of these electronic effects is thus given by the difference between the pK_a value of the substituted benzoic acid and that of benzoic acid itself. A parameter σ, the *substituent constant*, may then be defined by (1.1) for

1

the equilibrium [*1.1*], which is a measure of the size of such effects for a given substituent in this case. K is the equilibrium constant for the substituted benzoic acid and K_0 is the equilibrium constant for benzoic acid. Since pK_a values are affected by both solvent and temperature, we

$$\sigma_x = \log K - \log K_0 = -pK_a + (pK_a)_0 \qquad (1.1)$$

arbitrarily define σ as referring to aqueous solution at 25 °C. Table 1.1 gives the pK_a values for *m*-nitro-, *p*-nitro-, *m*-methyl-, and *p*-methyl-benzoic acids, and benzoic acid itself, from which by use of (1.1), σ_m and σ_p for these two groups have been calculated.

TABLE 1.1 pK_a *values of some benzoic acids* (H_2O, 25 °C)

X	pK_a (meta)	pK_a (para)	σ_m	σ_p
NO_2	3.50	3.43	0.71	0.78
CH_3	4.28	4.38	−0.07	−0.17

(The pK_a of benzoic acid is 4.21.)

Electron-withdrawing groups, such as NO_2, increase the equilibrium constant K of [*1.1*] because they stabilise the carboxylate anion; thus the pK_a is numerically smaller than that for benzoic acid, and σ is consequently *positive*. Electron-donating groups, CH_3 for example, decrease the equilibrium constant, and so their σ values are *negative*.

In table 1.2 are shown substituent constants calculated from (1.1) for a variety of common groups (Jaffé, 1953; McDaniel & Brown, 1958).

We must now examine whether such σ values are really constant; that is, do they correctly predict substituent effects on other side chain reactions or equilibria of benzenoid molecules? In many instances they do, and this is illustrated by the following two examples.

TABLE 1.2 *Substituent constants for common groups*

Substituent	σ_m	σ_p	Substituent	σ_m	σ_p
$N(CH_3)_2$	−0.21	−0.83	NH_2	−0.16	−0.66
OCH_3	0.12	−0.27	OCH_2CH_3	0.10	−0.24
CH_3	−0.07	−0.17	CH_2CH_3	−0.07	−0.15
$CH(CH_3)_2$	−0.07	−0.15	$C(CH_3)_3$	−0.10	−0.20
F	0.34	0.06	Cl	0.37	0.23
Br	0.39	0.23	I	0.35	0.28
$COOC_2H_5$	0.37	0.45	$COCH_3$	0.38	0.50
CN	0.56	0.66	NO_2	0.71	0.78
$\overset{+}{N}(CH_3)_3$	0.88	0.82			

The first is the acid dissociation of phenylphosphonic acids (Jaffé, Freedman & Doak, 1953):

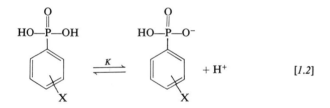

$$+ H^+ \qquad [1.2]$$

The second is the alkaline hydrolysis of ethyl benzoates (Ingold & Nathan, 1936; Evans, Gordon & Watson, 1937):

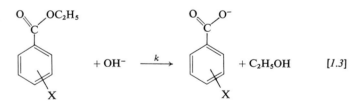

$$+ OH^- \xrightarrow{k} \qquad + C_2H_5OH \qquad [1.3]$$

Data for the two reactions are recorded in table 1.3, while figs. 1.1 and 1.2 show the graphs of log K/K_0 for the acid dissociations against σ, and log k/k_0 for the hydrolyses against σ, respectively. Both of these graphs show good linearity for *m*- and *p*-substituents. As fig. 1.2 reveals, however, *o*-substituents do not fall on, or even near, the line defined by the other substituents; this demonstrates that steric interactions between substituent and reaction centre, masking electronic factors, vary from

TABLE 1.3   *Selected equilibria and reaction data*
Acid dissociation constants of phenylphosphonic acids
(H_2O, 25 °C)

Substituent	pK_a	Substituent	pK_a
H	1.83	p-Br	1.60
m-NO_2	1.30	m-Cl	1.55
p-NO_2	1.24	p-Cl	1.66
m-Br	1.54	p-CH_3	1.98

Specific rate constants for alkaline hydrolysis of ethyl benzoates (85 % *aq. ethanol*, 25 °C)

Substituent	$10^5\,k/\mathrm{l\,mol^{-1}\,s^{-1}}$	Substituent	$10^5\,k/\mathrm{l\,mol^{-1}\,s^{-1}}$
o-$OCH_3{}^a$	7.8	p-Cl	267
p-OCH_3	13	o-$NO_2{}^a$	541
m-CH_3	43	m-NO_2	4290
p-CH_3	28	p-NO_2	7200
H	62.1	o-Cl^a	139
m-Cl	477		

^{a} The relevant values of σ_o ($\log K - \log K_o$) for the o-substituted benzoic acids are 0.29 for o-CH_3, 0.26 for o-Cl, and 1.03 for o-NO_2.

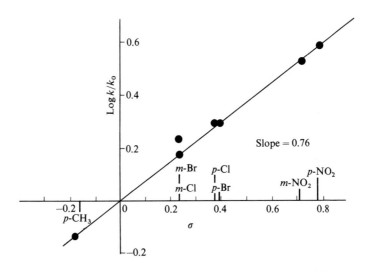

Fig. 1.1. Log K/K_0 *vs.* σ, dissociation of phenylphosphonic acids.

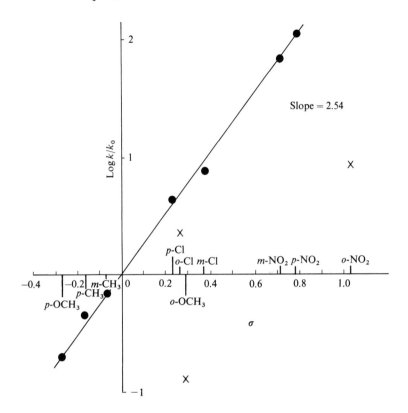

Fig. 1.2. Log k/k_0 *vs.* σ, alkaline hydrolysis of ethyl benzoates; σ_m and σ_p, ●; σ_o, ×.

one reaction type to another, and *o*-substituents must be left out of the reckoning, at least for the present.

The linearity of these plots for the remaining substituents reveals that the dissociation of benzoic acids, the dissociation of phenylphosphonic acids, and the alkaline hydrolysis of ethyl benzoates respond in the same manner to the influence of *m*- and *p*-substituents. The slope of the correlation line, 0.76 and 2.54, for the latter two respectively, is referred to as ρ, the *reaction constant*. It is a measure of the sensitivity of the reaction or equilibrium to electronic substituent effects, being by definition 1.00 for the dissociation of benzoic acids in water at 25 °C. In both of the examples chosen, ρ is also positive. This means, for the acid–anion equilibrium, that electron-donating groups decrease the extent of dissociation, while electron-withdrawing groups increase it, as can be readily argued from their effect on the stability of the anion. Similarly, electron donor substituents retard the rate of the hydrolysis reaction, while electron acceptors accelerate it; the carbon atom of the carbonyl group of an ester will become more reactive towards nucleophiles with electron-withdrawing groups present.

Generally, therefore, side chain reactivity of *m*- and *p*-substituted benzenes correlates linearly with σ values. For a reaction we can thus write the equation

$$\log k/k_0 = \sigma\rho \tag{1.2}$$

where k is the rate constant for the side chain reaction of a *m*- or *p*-substituted benzene derivative, and k_0 is the rate constant for the unsubstituted compound. The analogous equation for a side chain equilibrium will be

$$\log K/K_0 = \sigma\rho \tag{1.3}$$

Equations (1.2) and (1.3) are known collectively as the Hammett equation (Hammett, 1940) which, since it correlates energy changes in reactions or equilibria with those in benzoic acid dissociations, represents a *linear free energy relationship*. It is useful here to summarise the general sign convention for (1.2) and (1.3), discussed above for particular examples, which follows from the definition of the sign of σ from (1.1). *If ρ for a side chain reaction is positive, that reaction is accelerated by electron-withdrawing substituents, and slowed down by electron-donating substituents; if it is positive for an equilibrium, electron-withdrawing substituents favour the right-hand side of the equation defining K for the equilibrium, relative to electron-donating substituents which favour the left-hand side.* A negative ρ value obviously implies the opposite of this for reactions and equilibria.

It is worth noting that all acid dissociations, of general form

$$BH^{n+} \rightleftharpoons B^{(n-1)+} + H^+ \tag{1.4}$$

will have positive σ values whether n is positive, negative, or zero, because the base $B^{(n-1)+}$ will be stabilised by electron-withdrawing substituents relative to the acid BH^{n+}.

Alternative forms of (1.2) and (1.3) are

$$\log k = \sigma\rho + \log k_0 \tag{1.5}$$

and

$$\log K = \sigma\rho + \log K_0 \tag{1.6}$$

Plotting $\log k$ or $\log K$ against σ should give a straight line with intercept $\log k_0$ or $\log K_0$ on the ordinate. This procedure involves less work than plots of (1.2) and (1.3), as the reader can verify if the data of table 1.3 are plotted in this manner, and it will be employed here subsequently. Another expedient is to multiply rate constants by a uniform factor (e.g. 10^5 in table 1.3) to avoid negative logarithms; clearly this will neither alter the degree of correlation nor the value of ρ. Notice also that, if pK_a values are plotted directly, *the ρ value is the negative of the measured slope.*

Finally, before considering the significance of ρ and σ in detail, a brief note with regard to the general accuracy of the Hammett correlation is appropriate. For certain sets of data relating to reactivities of benzene side chains, deviations from the simple Hammett equation may be observed which are large, uniform, and confined to distinct classes of substituents in the *p*-position. Such deviations imply a breakdown in the basic assumptions of the equation; they are mechanistically significant, and will be considered later. At this stage, we are concerned rather with small random deviations, such as can be seen in figs. 1.1 and 1.2 where, although the correlation is good, there are some slight deviations from the line. These have their origin in either experimental inaccuracies, which are generally much larger than the particular workers concerned estimate them to be, or second order irregularities in behaviour of individual substituents. Indeed, the experimental values recorded for the pK_a's of substituted benzoic acids, fundamental to the evaluation of σ, vary quite considerably from one worker to the next, even though measured under the same conditions. This is illustrated very clearly, for example, in McDaniel & Brown's work (1958). Considerations of this sort lead to the conclusion that the third decimal place in σ or ρ is absolutely meaningless. Even deductions putting emphasis on the second place should be treated with some reservation.

1.3. The reaction constant ρ. Table 1.4 gives ρ values for a few reactions and equilibria. Later, references will be made to many further reaction constants. Those in the table serve as simple examples from which several general points can be made. Firstly, it should be noted that all the equilibria chosen are acid dissociations, and consequently the ρ values are positive.

The reaction constant ρ has been interpreted as a measure of the susceptibility of the reaction or equilibrium to substituent effects; a large ρ value for an equilibrium thus implies a large charge *change* at

TABLE 1.4 *Some typical reaction constants*

Equilibria		ρ
[1.4]	$XC_6H_4COOH \rightleftharpoons XC_6H_4COO^- + H^+$ (H$_2$O, 25 °C)	1.00
[1.5]	(50% aq. C$_2$H$_5$OH, 25 °C)	1.60
[1.6]	(C$_2$H$_5$OH, 25 °C)	1.96
[1.7]	$XC_6H_4CH_2COOH \rightleftharpoons XC_6H_4CH_2COO^- + H^+$ (H$_2$O, 25 °C)	0.49
[1.8]	$XC_6H_4CH_2CH_2COOH \rightleftharpoons XC_6H_4CH_2CH_2COO^- + H^+$	
	(H$_2$O, 25 °C)	0.21
[1.9]	$XC_6H_4CH{=}CHCOOH \rightleftharpoons XC_6H_4CH{=}CHOO^- + H^+$	
	(H$_2$O, 25 °C)	0.47
[1.10]	$XC_6H_4\overset{+}{N}H_3 \rightleftharpoons XC_6H_4NH_2 + H^+$ (H$_2$O, 25 °C)	2.77
[1.11]	(30% aq. C$_2$H$_5$OH, 25 °C)	3.44
[1.12]	$XC_6H_4OH \rightleftharpoons XC_6H_4O^- + H^+$ (H$_2$O, 25 °C)	2.11
[1.13]	$XC_6H_4PO(OH)_2 \rightleftharpoons XC_6H_4PO.OH.O^-$ (H$_2$O, 25 °C)	0.76
[1.14]	(50% aq. C$_2$H$_5$OH, 25 °C)	0.99
Reactions		
[1.15]	$XC_6H_4COOC_2H_5 + OH^- \rightleftharpoons XC_6H_4COO^- + C_2H_5OH$	
	(85% aq. C$_2$H$_5$OH, 25 °C)	2.54
[1.16]	$XC_6H_4CH_2OCOCH_3 + OH^- \rightleftharpoons XC_6H_4CH_2OH + CH_3COOH$	
	(60% aq. (CH$_3$)$_2$CO, 25 °C)	0.47
[1.17]	$XC_6H_4N(CH_3)_2 + CH_3I \rightleftharpoons XC_6H_4N^+(CH_3)_3I^-$	
	(90% aq. (CH$_3$)$_2$CO, 35 °C)	−3.30
[1.18]	$XC_6H_4NH_2 + C_6H_5COCl \rightleftharpoons XC_6H_4NHCOC_6H_5 + HCl$	
	(C$_6$H$_6$, 26 °C)	−2.78

the atom of the side chain attached directly to the ring carbon atom from one side of the equilibrium to another. Therefore interposition of a methylene unit between the side chain and the ring results in considerable decrease in ρ. This is illustrated by the sequence of dissociations of benzoic, phenylacetic and phenylpropionic acids [1.4], [1.7] and [1.8]. There is however an increase in ρ value from equilibrium [1.8] (phenylpropionic acid) to [1.9] (cinnamic acid). This is due to the additional mode of communication, in the latter case, of the substituent effect at the ring carbon atom to the reaction centre by conjugation through the double bond.

Varying the solvent will also affect the size of ρ. This is shown by sequences [1.4], [1.5] and [1.6], and [1.13] and [1.14]. As the solvent is changed from water (dielectric constant 79) to ethanol (dielectric constant 24) there is a decreasing degree of charge stabilisation by solvation of the acid anion relative to the neutral acid function. This results in greater susceptibility of the dissociation to substituent effects, and thus ρ increases.

When we turn our attention to the irreversible reactions [*1.15*], [*1.16*], [*1.17*] and [*1.18*], we must remember that such reactions can be interpreted as equilibria between the reactants and the transition state (TS) of the reaction

$$\text{Reactants} \xrightleftharpoons{K^{\ddagger}} \text{TS} \xrightarrow{k^{\ddagger}} \text{Products} \qquad (1.7)$$

$$K^{\ddagger} = [\text{TS}]/[\text{Reactants}] \qquad (1.8)$$

This notion is considered in some detail in chapter 5, but at this stage it must be noted that for interpretation of the relevant σ values, the charge difference between the ground state reactant molecule and the transition state (or the reaction intermediate, if one occurs, which for our purpose will be a close enough approximation) is considered. This is illustrated in detail for reactions [*1.15*] and [*1.17*] in fig. 1.3.

When the formation of the transition state involves electron capture by the reaction site, acceleration is produced by electron acceptor substituents and ρ is positive, e.g. [*1.15*] and [*1.16*].

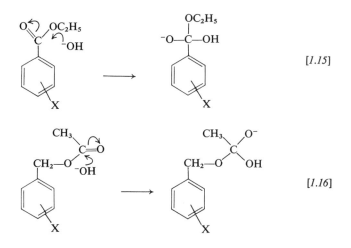

[*1.15*]

[*1.16*]

Alternatively, when formation of the transition state involves donation of electrons from the reaction site, the reaction is accelerated by electron donors, and ρ is consequently negative, e.g. [*1.17*] and [*1.18*].

The size of ρ is an indication of the extent of charge development at the atom of the reacting side chain adjacent to the ring in passing from ground to transition state. In three of the examples shown ([*1.15*], [*1.17*], [*1.18*]) this atom is directly involved in the bond making and

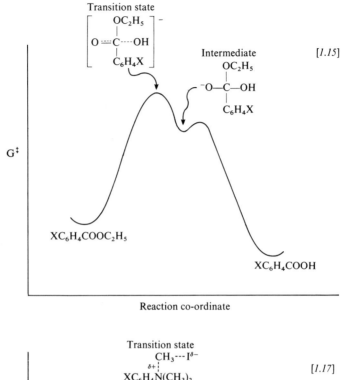

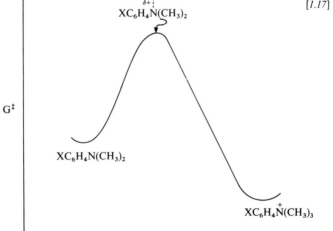

Fig. 1.3. Free energy profiles for reactions [*1.15*] and [*1.17*].

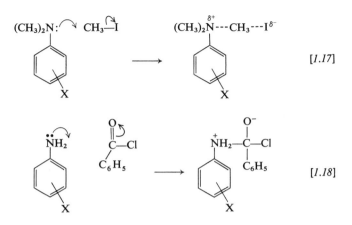

breaking process, and ρ is consequently relatively large. On the other hand, the reaction site in the benzyl acetate hydrolysis [*1.16*] is insulated from the ring by a CH_2 group and ρ is smaller.

The question of the effect of temperature on ρ values is taken up in chapter 5.

1.4. The substituent constant σ. A substituent attached to the benzene nucleus will affect the electronic distribution within that nucleus. σ values represent the measurement of this charge distribution by attachment of a reacting side chain to the *m*- and *p*-positions, in much the same way that one would measure electrical potential at a given point in a macroelectrical system by attachment of a voltmeter at that position.

We will assume, as foreshadowed in the previous discussion, that the electronic influence of a substituent as expressed by its σ_m and σ_p values can be quantitatively divided into the sum of independent inductive and resonance contributions. The inductive effect I of the substituent originates in the polarisation of the electrons in the σ bond connecting it to the ring. It is due to the effective electron attracting or repelling property of the substituent relative to that of an sp^2 carbon atom of a benzene ring. Since the orbital overlap in σ bonds is along the axes of the orbitals, it is efficient and the resultant bond is strong. The electrons of such a bond, and thus the inductive effect to which they give rise, are therefore of low polarisability. This means that the inductive effect is constant; its variation on perturbation of the benzene ring by attachment of other groups or by approach of a reagent molecule is small. The effect of this primary induction on the benzene ring can occur in three different ways. These are as follows: by direct electrostatic interaction through space, which is called the direct or electrical field effect, symbol D: by successive polarisations of adjoining σ bonds in the σ framework of the benzene ring, symbol I_σ: by interaction with the aromatic π system, symbol I_π.

The experimental recognition, separation and evaluation of these secondary effects are problems currently receiving much attention. Several points bearing on the question will occur later in the text. For the present it may be noted that I_π appears to be small. This result might be anticipated because the σ bond joining the substituent to the ring is orthogonal to the aromatic π bonds and therefore interaction is minimal. A simple practical observation to support this is the coincidence of the ultraviolet (UV) spectra of benzene and trimethylanilinium ion (Bishop & Craig, 1963). Perturbation of the π cloud strongly influences UV spectral patterns. The conclusion seems to be that interaction between the π electrons and the powerful inductive effect of $\overset{+}{N}(CH_3)_3$ is absent, although this deduction is not a rigorous one, because it is possible that the ground state orbitals and those to which the electrons are promoted in the observation of the UV spectrum are influenced by the substituent, but to an equal extent. Usually the field effect is considered independently of I_σ, but both are 'classical' interactions dying away with increasing distance between origin and point of interaction and are thus very hard to differentiate experimentally. It seems logical at this stage therefore to group them together, but it is important to realise that if both modes of propagation are important, then the *ratio* of the two for any given substituent from both the *m*- and *p*-position must be constant for all reactions obeying the Hammett equation.

To summarise, we may write for the influence of the inductive effect of a substituent on the *m*- and *p*-positions of the benzene ring:

$$I(\text{total}) = D + I_\sigma + I_\pi \qquad (1.9)$$

$$I_\pi \simeq 0 \qquad (1.10)$$

Therefore

$$I(\text{total}) \simeq D + I_\sigma \qquad (1.11)$$

The resonance effect R of the substituent involves interaction between orbitals in the substituent which are in the same plane as the π electron orbitals of the benzene ring. Such a 'non-classical' or quantum interaction affects alternate positions and is thus registered at positions o and p to the point of attachment of the substituent giving rise to it.

This type of bond interaction is depicted in acetophenone (**1**). Here

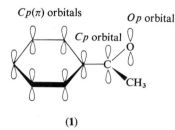

(1)

bond formation involves overlap parallel to the axes of the orbitals. This has two important consequences.

Firstly, such overlap is not efficient; resonance effects are of *high polarisability*. They are sensitive to the influence of other groups introduced into the benzene ring, or to the approach of a reagent.

Secondly, twisting the group out of the plane of the ring, for example by introducing two flanking methyl groups into the adjacent *o*-positions, reduces this type of overlap and in consequence the resonance interaction. Notice however that the inductive effect of the group would remain unchanged.

Thus, taking into consideration the assumption that the inductive effect is transmitted about equally to the *m*- and *p*-positions (the justification for which is given below), σ_m is an approximate measure of the size of the inductive effect, and $(\sigma_p - \sigma_m)$ an approximate measure of the resonance effect, of a given substituent. Writing σ_I for that portion of σ due to the inductive effect, and σ_R for that due to resonance,

$$\sigma_m = \sigma_I \tag{1.12}$$

$$\sigma_p = \sigma_I + \sigma_R \tag{1.13}$$

$\overset{+}{N}(CH_3)_3$ cannot exert either a large $+R$ or $-R$ effect, a fact we have already used in arguing for the magnitude of I_π. The σ values for this group (table 1.2) thus yield approximate information on the I effect alone. Since the substituent bears a positive charge on the atom linking it to the ring, the σ values are large and positive, indicating powerful electron withdrawal. The fact that σ_m is only slightly larger than σ_p tells that the inductive effect of a substituent has substantially the same influence at both positions.

Jaffé (1953) has reached this conclusion by a different approach. Table 1.4 gives the ρ value for ionisation of cinnamic acids as 0.47.

Therefore each sp^2 carbon atom in the side chain suppresses the substituent effect by a factor of $\sqrt{0.47}$ or 0.69. If we assume that this factor is applicable to sp^2 atoms within the benzene ring, the ratio of inductive transmission *via* the σ bonds to the *m*- (2) and *p*-position (3) is 1.47/1.38

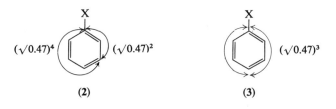

(2)   (3)

which is 1.1. With this value we may compare σ_m/σ_p for $\overset{+}{N}(CH_3)_3$ which is 1.07. Of course such a model assumes that the inductive effect is conveyed entirely by the σ bonds. The ratio of the distances through space between the point of attachment of a substituent and the *m*- and *p*-position is 0.85. This indicates that the ratio σ_m/σ_p for an inductive substituent assuming a direct effect as the mode of transmission would be 1.2. Either method of transmission, or a combination of both, are thus compatible with these approximate calculations.

Groups such as CHO, $COCH_3$, $COOC_2H_5$, CN and NO_2 are of dipolar type, with the positive end of the dipole attached to the benzene ring. They exert their influence on the ring by a combination of $-I$ and $-R$ effects. The former effect outweighs the latter, as the following figures taken from table 1.2 demonstrate:

	σ_m (σ_I)	$\sigma_p - \sigma_m$ (σ_R)
$COOC_2H_5$	0.37	0.08
$COCH_3$	0.38	0.12
CN	0.56	0.10
NO_2	0.71	0.07

It will be noticed there that the generally accepted sign convention for inductive and resonance effects, $+$ for electron donation and $-$ for electron withdrawal, is the opposite to that for σ values. This may initially cause some confusion, but both conventions are now too well established to allow a change. Bowden & Shaw (1971) have reported the measurement of σ_p values of 0.48, 0.47 and 0.32 for the series of substituents $COCH_2CH_3$, $COCH(CH_3)_2$ and $COC(CH_3)_3$ respectively. For the first two members of the series, the σ_p values are essentially the same as that for $COCH_3$. Changing the character of the alkyl group will not alter significantly the overall inductive effect of these substituents, while the carbonyl function remains in the plane of the ring because the methyl groups can twist away from the *o*-hydrogens. (Reference to structure (1) may be helpful here.) However, when a third CH_3 is introduced steric interaction with the *o*-hydrogens becomes large. The carbonyl group is therefore twisted out of the plane of the ring as in (4) with a loss of resonance interaction indicated by the reduced positivity of the σ_p value.

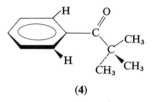

(4)

The remaining substituents in table 1.2, with the exception of alkyl, fall into the $-I +R$ category. The halogens have a powerful $-I$ effect due to their electronegativity which more than compensates $+R$, arising from the donation of the halogen lone pair electrons into the ring.

	σ_m (σ_I)	$\sigma_p - \sigma_m$ (σ_R)
F	0.34	−0.28
Cl	0.37	−0.14
Br	0.39	−0.16
I	0.35	−0.07

For the other groups in table 1.2 of $-I$ and $+R$ class, the resonance effect into the ring is greater than inductive withdrawal. As the resonance effect gets larger, however, more caution must be applied to use of the assumption that σ_m reflects inductive effects alone, because R may be felt at the *m*-position, although in very much diminished form. This secondary effect presumably arises by inductive transmission of the resonance perturbation at the *o*- and *p*-position to the *m*-position (5), (6), (7). The positive σ_m value for OCH_3 shows that the inductive effect is still sufficiently powerful relative to the resonance effect to predominate at the *m*-position. However, the negative values of σ_m for NH_2 and $N(CH_3)_2$ indicates that in these instances the very strong $+R$ effect does have a proportionately larger influence than the $-I$ effect even at the *m*-position. One would expect, however, OCH_3 to have a larger inductive effect than NH_2 or $N(CH_3)_2$ because of the greater electronegativity of the oxygen compared with the nitrogen atom.

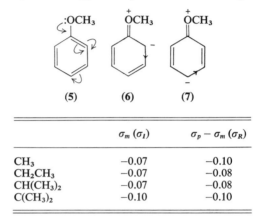

	$\sigma_m\ (\sigma_I)$	$\sigma_p - \sigma_m\ (\sigma_R)$
CH_3	−0.07	−0.10
CH_2CH_3	−0.07	−0.08
$CH(CH_3)_2$	−0.07	−0.08
$C(CH_3)_2$	−0.10	−0.10

Alkyl substituents are the only common substituents with a $+I$ effect. Since the σ_p values are more negative than the σ_m values there must also be a contribution from a $+R$ effect. This must arise by hyperconjugation, $(8) \leftrightarrow (9)$. The values for $C(CH_3)_3$ show that carbon hyperconjugation

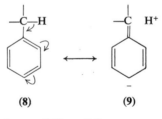

is also an important factor, $(10) \leftrightarrow (11)$.

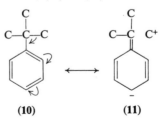

Notice that an identical effect can be postulated for $\overset{+}{N}(CH_3)_3$ as in **(12)**.

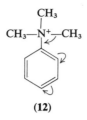

(12)

We must therefore be cautious about treating this latter substituent entirely as of $-I$ character, although it can certainly be regarded as having predominantly this effect.

The analysis applicable to substituents by consideration of their σ values is thus plain. Even if not completely unambiguous, this is a considerable improvement on the completely qualitative description. This type of analysis can be extended and refined by application of Hammett type equations to systems specifically designed to measure certain effects and eliminate others. Experiments and calculations of this kind will be discussed later, particularly in chapter 3.

1.5. The σ^0 scale. The function of the reacting side chain is to measure the electronic interaction between substituent and ring. In many cases, however, the substituent may be able to resonate directly with the side chain. This is known as *through conjugation*. In fact, such conjugation is possible in the equilibrium used to define σ values [*1.1*]. Structures **(13)** and **(14)**, and **(15)** and **(16)** display this for a resonance donor in benzoic acid and benzoate anion, respectively.

What contribution does such a resonance make to σ values? The answer to this question was sought by several workers, mentioned later. Side chain reactions were studied in which through conjugation was absent, the usual device being insulation of the side chain from the ring by introduction of one or two methylene units. From this work new σ values, called σ^0 values, were found.

Before describing the scale in greater depth, it is *very important* to notice that through conjugation with the reacting side chain is not in itself responsible for differences between σ and σ^0 values. The need for modified σ values of any kind arises from a *change* of through conjugation between one side of a thermodynamic equilibrium and the other, or between the ground state and the transition state for the kinetic equilibrium of a reaction (1.7). If it occurred to the same extent on both sides

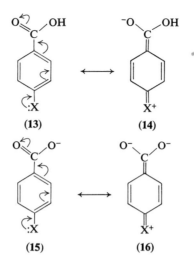

(13)   (14)

(15)   (16)

of the equilibrium, it would not alter the position of the equilibrium. Hence it would not figure in K or $K^\dagger$ and would make no contribution to the magnitude of substituent effects.

For all *m*-substituents, σ and σ^0 are equivalent, since the deviations which do occur are due to resonance effects, and so do not arise appreciably with groups in the *m*-position. *p*-Substituents also have the same σ and σ^0 values, with the exception of resonance donors, and then only if such resonance is strong. OCH_3, NH_2, and $N(CH_3)_2$ form the commonest examples and values for these substituents are shown in table 1.5. The $(\sigma_p - \sigma^0_p)$ values show the extent of through conjugation present in the benzoic acid, (13) ↔ (14) which is suppressed in the anion, (15) ↔ (16) owing to alternative carboxylate anion resonance. Figures 1.4 and 1.5 demonstrate the correlation for dissociation of phenylacetic acids and saponification of ethyl phenylacetates. These graphs reveal one serious drawback to the evaluation of accurate σ^0 values: the necessity for the

TABLE 1.5 σ^0 *values*

Substituent	σ_p	σ^0_p	$\sigma_p - \sigma^0_p$
N(CH₃)₂	−0.83	−0.44	−0.39
NH₂	−0.66	−0.38	−0.28
OCH₃	−0.27	−0.12	−0.15

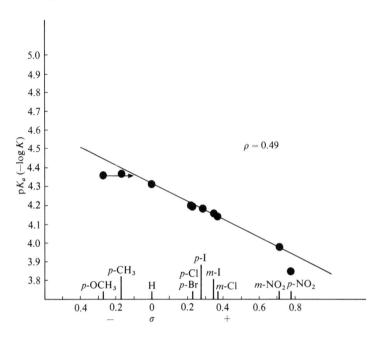

Fig. 1.4. Log K *vs.* σ for dissociation of phenylacetic acids (H_2O, 25 °C).

insulating methylene linkages reduces the sensitivity of the side chain to substituent effects as shown by the low ρ values, and the correlations show much random scatter.

The origin of σ^0 values and the nature of their deviation from σ values is conceptually quite rational, but as a consequence of historical development complexities arise. Van Bekkum, Verkade & Wepster (1959) first suggested the need for a scale in which the effects of through conjugation were eliminated. They derived values from quite a wide selection of reactions, and these values were called σ^n. Taft (1960) used only four reactions. These were the ionisations of $XC_6H_4CH_2COOH$ and $XC_6H_4CH_2CH_2COOH$, and alkaline hydrolysis of $XC_6H_4CH_2COOC_2H_5$ and $XC_6H_4CH_2OCOCH_3$ (the latter given in table 1.4) for which he introduced the σ^0 terminology. Norman *et al.* (1961) developed σ^G values from the rates of alkaline hydrolysis of $XC_6H_4CH_2COOC_2H_5$. Finally, Yukawa, Tsuno & Sawada (1966) confirmed the σ^0 correlation with further measurements of hydrolyses of $XC_6H_4CH_2COOC_2H_5$. For all practical purposes, however, σ^n, σ^0 and σ^G are the same, and equal to σ apart from the values of strong resonance donors in the *p*-position.

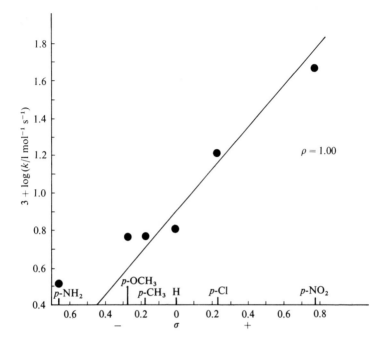

Fig. 1.5. Log k *vs.* σ, basic hydrolysis of ethyl phenylacetates (88% aq. ethanol, 30 °C).

Judging from its theoretical basis, the σ^0 scale, with its absence of any direct resonance interactions between side chain and substituents, appears to be more fundamental than the σ scale. Nevertheless, when the general correspondence between σ^0 and σ for the great majority of substituents, the low precision of the former values, and the convenient and simple experimental basis, the ionisation of benzoic acids, for definition of the σ scale are all taken into account, a good case can be made for retaining this latter scale in a primary role. At the same time any departure of p-amino or p-alkoxy substituents from a correlation using σ values which indicates they are exerting a diminished electron-donating effect could well indicate a reaction series which is appropriately considered in the terms used to warrant the σ^0 scale.

1.6. The effect of solvent on σ values. Solvent has a large effect on the magnitude of ρ. However, it appears to have little influence generally on σ values, judged within the context of the accuracy of the Hammett equation. A good illustration of this is afforded by the work of Kondo,

Matsui & Tokura (1969). Look at the data in fig. 1.6. The reaction involved is the quaternisation of substituted dimethylanilines with methyl picrate.

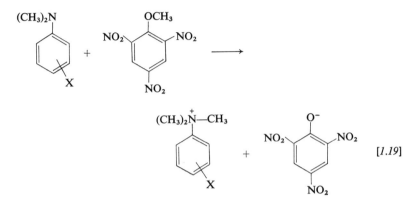

[*1.19*]

The type of reasoning explained in § 1.3 leads to the conclusion that the ρ value should be large and negative; the reaction rate will be very sensitive to substituent effects. For the two reaction solvents shown, benzene and acetonitrile, there is some difference in ρ (−2.72 and −2.37 respectively), as well as in the rate constants of the unsubstituted dimethylaniline (0.002 and 0.01 l mol^{-1} s^{-1} respectively). This reflects the difference in solvation properties of the solvents, the polar cyanide molecules stabilising the polar transition state more efficiently than the hydrocarbon. But the correlation with σ values in both cases is excellent. These workers also studied the reaction in a number of other solvents, p-xylene, nitrobenzene, bromobenzene, methanol, ethyl methyl ketone, tetrahydrofuran and chloroform, and although ρ continued to vary, the correlation was in all cases highly accurate.

In a striking confirmation and extension of this conclusion, it is found that the rates of vapour phase reactions, in the complete absence of solvent, also correlate with the Hammett equation. Examples of this are found in the extensive studies of G. G. Smith who has examined the pyrolysis of various esters and carbonates. Fig. 1.7 shows the graph obtained for substituted ethyl phenyl carbonates, which are considered to decompose via a unimolecular *cis* elimination, yielding the primary products shown in [*1.20*] (Smith & Kelly, 1971).

The correlation is remarkably good, considering that not only is solvent absent, but also that the temperature is some 384 °C, 359 °C in

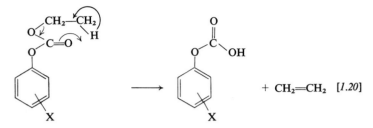

$$+ \; CH_2\!\!=\!\!CH_2 \quad [1.20]$$

excess of that for the definition of σ, and that the reaction is very insensitive to substituent effects ($\rho = 0.19$). Indeed, the accuracy is sufficiently high to disclose that σ^0 values should be used for p-NH_2 and p-OCH_3,

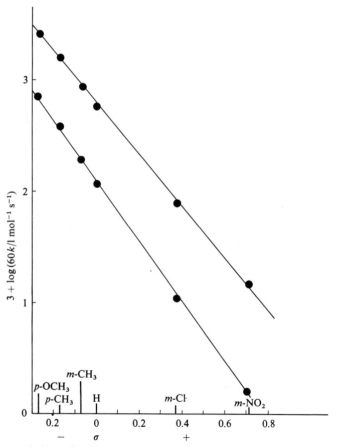

Fig. 1.6. Solvent effects on the quaternisation of dimethylanilines with methyl picrate.

as expected from the presence of an insulating oxygen atom between the aromatic ring and reacting side chain.

There are certain substituents which are capable of entering into such pronounced solvent interactions as significantly to alter their structure and therefore their electronic character. Thus the highly acidic hydrogen atom of the OH group will form strong hydrogen bonds with hydrogen bond acceptor solvents and the group in certain media will function as intermediate in electronic character between OH and O^-, behaviour reflected in its effective σ value.

In acidic media, the NH_2 and $N(CH_3)_2$ groups may bond to a proton, so that determination of their σ values can be complicated by zwitterion formation (McDaniel & Brown, 1958) as shown in the scheme below:

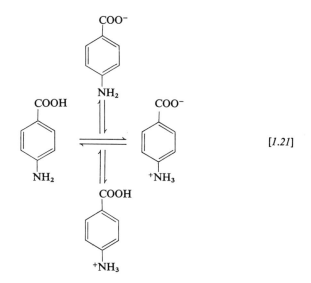

[1.21]

This type of system is capable of only approximate elucidation by means of model compounds, a method explained for the analogous case of pyridine carboxylic acids in chapter 4. Such complications detract from the confidence which may be placed on the σ values quoted for NH_2 and $N(CH_3)_2$ in table 1.2. This appears to be particularly unfortunate, because such strong electron donors extend the σ scale on the negative side to an extent which cannot be realised by any other common group, but of course the facile co-ordination of the amino lone pair electrons with acidic species is a manifestation of the electronic properties

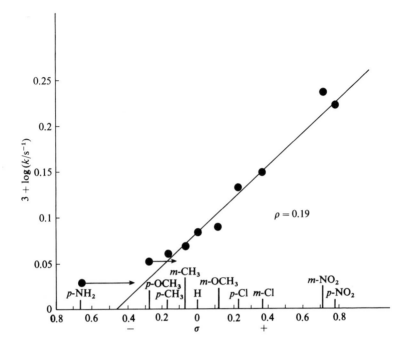

Fig. 1.7. Vapour phase decomposition of ethyl phenyl carbonates.

of the group which in alternative circumstances leads to its uniquely negative σ values.

However, the general conclusion must be, that although solvation will certainly be one of the main factors contributing to the deviations from precise correlation which occur in all Hammett plots, in most cases it has only a small effect on the nature and extent of substituent interaction with the benzene ring. This conclusion is a necessary one. If solvents did have a large effect on σ values, each substituent would require different σ_m and σ_p values for each solvent in which reactions were studied. The Hammett equation would then become unmanageably complicated.

1.7. Problems

1. The relative rates of alkaline hydrolysis of substituted benzamides in water at 100 °C are as follows (Reid, 1900):

Substituent	Rel. rate	Substituent	Rel. rate	Substituent	Rel. rate
m-I	2.60	*m*-NO$_2$	5.60	*p*-OCH$_3$	0.49
p-I	1.69	H	1.00	*m*-NH$_2$	0.93
m-Br	2.97	*m*-CH$_3$	0.83	*p*-NH$_2$	0.20
p-Br	1.91	*p*-CH$_3$	0.65	*m*-OH	0.19

Demonstrate the applicability of the Hammett equation to this reaction, calculate the ρ value, and comment on any deviations from the correlation.

2. Interpret the following σ values in terms of the electronic character of each group:

Substituent	σ_m	σ_p	
NHCOCH$_3$	0.21	0.00	
C$_6$H$_5$	0.06	−0.01	
CF$_3$	0.43	0.54	
CH=CH—NO$_2$	0.32	0.26	(Stewart & Walker, 1957)
C$_3$H$_5$ (cyclopropyl)	−0.07	−0.21	(Šmejkal, Jonáš & Farkaš, 1964)
SOCH$_3$	0.52	0.49	
SO$_2$CH$_3$	0.60	0.72	
OCF$_3$	0.40	0.35	(Sheppard, 1963)

All the values are taken from McDaniel & Brown (1958) unless otherwise specified.

3. The σ_m and σ_p values for the formyl group have been estimated as 0.36 and 0.44 respectively by measurement of the pK_a values of *m*- and *p*-formylbenzoic acids (Humffray, Ryan, Warren & Yung, 1965; Bowden & Shaw, 1971), and as 0.35 and 0.22 respectively by measurement of the rate of alkaline hydrolysis of *m*- and *p*-formyl ethyl benzoates in 56 % aqueous acetone (Tommila, 1941). How may the discrepancy be accounted for?

4. The picryl substituent, $(NO_2)_3C_6H_2$, has values of σ_m 0.43, σ_p 0.41 (Ruskie & Kaplan, 1965). What conclusions may be drawn from these figures regarding the relative size of the inductive effect transmitted from the *m*-position as compared with that from the *p*-position, and the configuration of the two benzene rings in this system?

5. The pK_a values of *m*- and *p*-monosubstituted benzoic acids in 50 %
aqueous ethanol correlate with σ with a ρ value of 1.60. The pK_a of
benzoic acid in this system is 5.71. The pK_a values of some 4-X,3,5-
dimethylbenzoic acids in this medium are shown below.

X	pK_a	X	pK_a
$N(CH_3)_2$	6.23	CN	4.90
NH_2	6.88	$COOCH_3$	5.44
Cl	5.59	NO_2	4.91
Br	5.55		

(Schaefer & Miraglia, 1964)

Use these results to examine and comment on the applicability of
additivity of σ values.

2 Elucidation of reaction mechanisms

2.1. Introduction. The systems regularly encountered in organic chemistry are of great complexity. The methods fashioned to account for them, such as the Hammett equation, molecular orbital calculations, solvent and isotope studies, thermodynamic and kinetic acidity functions and the Brønsted equation, are thus necessarily empirical approximations. An argument can be made for the individuality of molecular orbital procedures, in that here one has commenced with theory, to which experimental data have been applied, while the others have involved initial establishment of empirical laws, to which theory is subsequently attached. But in the final analysis, it must be admitted that all involve a blend of theory and empiricism in very similar proportions. Consequently, when each approach is viewed in isolation, it seems at best vulnerable to criticism, and at worst hopelessly naive. However, considerable encouragement is gained from the fact that application of one method to a given problem frequently, if not invariably, corroborates the conclusions reached concerning that system using alternative and essentially independent approaches. Progress in the understanding of mechanistic organic chemistry is achieved, not by setting up one method in competition with others, from which there would seem little to be gained, but by the union of all techniques, employing as many as possible in a combined assault on the problem in hand. It is well beyond the scope of this book to treat all such methods in detail, but the student is reminded of the relevance of this multiform approach and in this and subsequent chapters the way in which use of the Hammett equation is supported by alternative procedures is briefly indicated. In particular, chapter 4 contains an account of confirmation and extension of conclusions reached with the use of the Hammett equation by means of simple molecular orbital methods.

2.2. Modified substituent constants. The distinction between σ and σ^0 values illustrated the general principle that resonance interactions between substituent and side chain may cause deviation from a single

27

unique equation, necessitating the use of modified substituent constants. It must be stressed again that such resonance interactions in themselves are not responsible for this – it is a result of a change in through conjugation, due to the high polarisability of the π electron system. Therefore, the need for modified substituent constants to correlate the observed rates for a reaction may indicate the resonance interactions in the transition state, then the form of the transition state itself, and in turn the nature of the reaction in which it is involved. Such modified substituent constants will now be considered, and shown to form a significant key to understanding reaction mechanisms.

2.3. The σ^- substituent constant. The ρ value for dissociation of anilinium ions in water at 25 °C is 2.77 from table 1.4, [*1.10*]. Using this value to calculate the log ratio of ionisation constants of *p*-nitroanilinium and unsubstituted anilinium ion from (1.3), we obtain $2.77 \times 0.78 = 2.16$. However, the observed value is 3.52, a discrepancy well outside experimental error, and implying an effective σ value of 1.27 for *p*-NO_2 in this system.

It is conjectured that this is due to through conjugation (**17**) which is not present in the nitroanilinium ion (**18**). This confers extra stability on the free base molecule, reducing its basicity. In accordance with this suggestion, the ordinary σ value for *m*-NO_2 functions quite accurately for prediction of the acidity of *m*-nitroanilinium ion; here NO_2 is not in conjugation with the amino group, and accordingly it exerts its influence predominantly through its inductive effect, which we have reasoned (§ 1.4) to be of low polarisability.

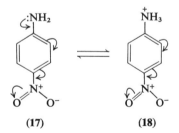

(17)   (18)

Table 2.1 gives pK_a values for a series of *m*- and *p*-substituted anilines, taken mainly from the work of Biggs & Robinson (1961). Plotting these against σ yields the graph shown in fig. 2.1.

All the *m*-substituents fall on or near the correlation line, together with all the *p*-substituents except those with a $-R$ effect capable of

TABLE 2.1   pK_a *values of substituted anilines* (H_2O, 25 °C)

Substituent	pK_a	Substituent	pK_a
H	4.58	m-CH$_3$	4.73
m-NO$_2$	2.50	p-CH$_3$	5.08
p-NO$_2$	1.06	m-OCH$_3$	4.23
m-COCH$_3$	3.55	p-OCH$_3$	5.34
p-COCH$_3$	2.21	m-NH$_2$	4.88
m-CN	2.76	p-NH$_2$	6.08
p-CN	1.76		
m-Cl	3.52		
p-Cl	3.98		

interaction as in (**17**). The exalted σ values necessary to correlate the latter substituents are called σ⁻ values, and examples for a range of common substituents are given in table 2.2.

TABLE 2.2   σ⁻ *substituent constants*

Substituent	σ	σ⁻	(σ⁻ − σ)
CHO	0.43	1.03	0.60
COOC$_2$H$_5$	0.45	0.68	0.23
COCH$_3$	0.50	0.84	0.34
CN	0.66	0.88	0.22
CF$_3$	0.54	0.74	0.20
NO$_2$	0.78	1.27	0.49

The correlation of the pK_a values of phenols is also predicted to fit σ⁻, not for the reason that there is no through conjugation in phenols, but because there would appear to be much greater through conjugation in the phenolate anion (**19** ⇌ **20**), where the oxygen atom bears a formal negative charge.

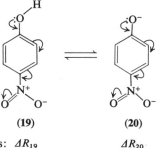

(**19**)   (**20**)

Through conjugations: ΔR_{19} ΔR_{20} $\Delta R_{19} < \Delta R_{20}$

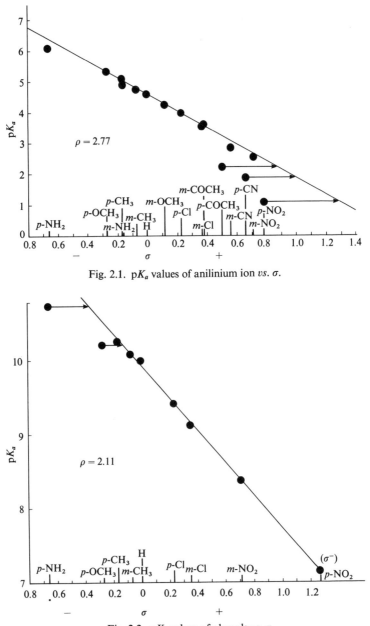

Fig. 2.1. pK_a values of anilinium ion *vs.* σ.

Fig. 2.2. pK_a values of phenols *vs.* σ.

The relevant plot is shown in fig. 2.2, for which the ρ value, 2.11, is given in table 1.4. The graph also shows that OCH_3 and NH_2 require σ^0 for correlation. Although there is no insulating group between OH and the ring, the resonance donation of this side chain into the ring in both its neutral and ionised forms is sufficient to suppress the alternative resonance of strong electron donor substituents, to the proportions indicated by σ^0. Notice that reactions [*1.17*] and [*1.18*] in table 1.4 would also correlate with σ^-.

2.4. The σ^+ substituent constant. By extrapolation of the argument given in the previous section, it might be expected that side chain reactions involving formation of an electron-deficient centre in direct conjugation with the benzene nucleus would require substituents of $+R$ type to display enhanced σ values, which could be termed σ^+ constants. This is indeed found to be the case. The reaction employed for determining such σ^+ constants is the S_N1 hydrolysis of substituted phenyldimethyl-carbinyl chlorides in 90% aqueous acetone at 25 °C (Stock & Brown, 1963).

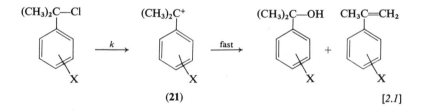

(21)   [2.1]

The transition state for the slow, rate-determining step must resemble the carbonium ion intermediate (21); substituents such as OCH_3 will therefore induce through conjugation (22) not present in the reactant molecule.

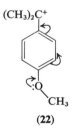

(22)

TABLE 2.3   *Rate constants for solvolysis of phenyldimethylcarbinyl chlorides (90% aq. acetone, 25 °C)*

Substituent	$10^5\,k/s^{-1}$	Substituent	$10^5\,k/s^{-1}$	Substituent	$10^5\,k/s^{-1}$
H	12.4	*m*-Cl	0.194	*m*-CN	0.0347
m-OCH₃	7.56	*p*-Cl	3.78	*p*-CN	0.0126
p-OCH₃	41 700	*m*-Br	0.178	*m*-NO₂	0.0108
m-CH₃	24.8	*p*-Br	2.58	*p*-NO₂	0.00319
p-CH₃	322	*m*-COOC₂H₅	0.269		
		p-COOC₂H₅	0.0806		

Table 2.3 gives values for selected rate constants of this reaction, while fig. 2.3 shows the plot of log k against σ.

The σ^+ constants are then defined in the following way. The reaction constant ρ is calculated, using σ_m values only, as -4.54, a large negative

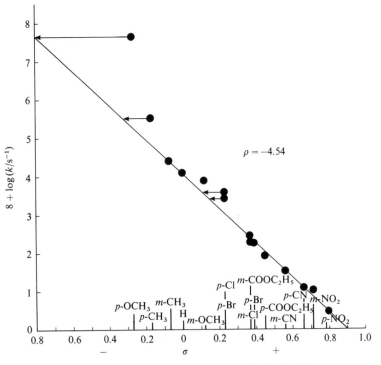

Fig. 2.3. Solvolysis of phenyldimethylcarbinyl chlorides.

TABLE 2.4 σ^+ *substituent constants*

Substituent	σ_m	σ^+_m	$(\sigma_m^+ - \sigma_m)$	σ_p	σ^+_p	$(\sigma^+_p - \sigma_p)$
$N(CH_3)_2$				−0.83	−1.7	−0.87
NH_2	−0.16	−0.16	0.00	−0.66	−1.3	−0.64
OCH_3	0.12	0.05	−0.07	−0.27	−0.78	−0.51
CH_3	−0.07	−0.07	0.00	−0.17	−0.31	−0.14
$C(CH_3)_3$	−0.10	−0.06	0.04	−0.20	−0.26	−0.06
C_6H_5	0.06	0.10	0.04	−0.01	−0.18	−0.17
F	0.34	0.35	0.01	0.06	−0.07	−0.13
Cl	0.37	0.40	0.03	0.23	0.11	−0.12
Br	0.39	0.41	0.02	0.23	0.15	−0.08
I	0.35	0.36	0.01	0.28	0.14	−0.14
$COOC_2H_5$	0.37	0.37	0.00	0.45	0.48	0.03
CN	0.56	0.56	0.00	0.66	0.66	0.00
NO_2	0.71	0.67	−0.04	0.78	0.79	0.01
$\overset{+}{N}(CH_3)_3$	0.88	0.36	−0.52	0.82	0.41	−0.41

value in line with development of extensive positive charge in the transition state, adjacent to the ring. The σ^+ values are then calculated from equation (2.1).

$$\log k/k_0 = -4.54\,\sigma^+ \qquad (2.1)$$

A selection of the values thus obtained are recorded in table 2.4.

The procedure for their definition means that they are different from σ values in nearly all cases. However, as the columns $(\sigma^+_m - \sigma_m)$ and $(\sigma^+_p - \sigma_p)$ show, these differences are very small and of no practical significance, with the exception, as expected, of those for p-substituents with a $+R$ component in their electronic composition.

The one errant substituent is $\overset{+}{N}(CH_3)_3$, which is considerably less deactivating in both m- and p-positions than expected. In the event this behaviour is quite inexplicable; presumably it will be the theme of further investigation.

2.5. Diagnosis of reaction mechanisms. So far our knowledge of relatively simple reactions and equilibria has been employed to make general deductions on the sign and size of ρ and σ, and to argue the necessity for definition of σ^0, σ^- and σ^+. This process can now be reversed – the ρ value and the type of σ value, necessary for correlation of substituent effects through a benzene ring on the rate of an unknown side chain reaction, can be used to throw light on the nature of the mechanism of that reaction. It is this type of procedure which has made the Hammett equation and its modifications of such great value.

Before considering some specific reaction types, it is important to notice that an overall reaction may consist of several different steps, so that the observed rate constant must be elucidated in terms of the equilibria and rate constants of these components. A very commonly encountered case is that in which an equilibrium between substrate S and reagent Y to form a reactive intermediate SY is succeeded by an irreversible step involving decomposition of SY to yield the reaction product P

$$S + Y \; \underset{k_{-1}}{\overset{k_1}{\rightleftharpoons}} \; SY \; \xrightarrow{k_2} \; P \qquad (2.2)$$

By applying the steady state approximation to (2.2):

$$k_1[S][Y] = k_{-1}[SY] + k_2[SY]$$

Therefore

$$[SY] = \frac{k_1[S][Y]}{k_{-1} + k_2}$$

Taking the rate of reaction to be that of formation of P:

$$\frac{d[P]}{dt} = k_2[SY] = \frac{k_1 k_2[S][Y]}{k_{-1} + k_2} \qquad (2.3)$$

If $k_{-1} \ll k_2$, (2.3) becomes

$$\frac{d[P]}{dt} = k_1[S][Y]$$

In this case, the free energy profile for the formation of the intermediate SY will be represented by fig. 2.4(a).

Writing the observed rate constant as k_{obs}:

$$\frac{d[P]}{dt} = k_{obs}[S][Y] = k_1[S][Y]$$

so that

$$k_{obs} = k_1 \qquad (2.4)$$

and if S represents a series of substituted benzene derivatives undergoing a side chain reaction, the correlation of k_{obs} with a Hammett-type equation applies directly to the single step process

$$S + Y \xrightarrow{k_1(k_{obs})} SY \tag{2.5}$$

Alternatively, if $k_{-1} \gg k_2$, (2.3) becomes

$$\frac{d[P]}{dt} = \frac{k_1 k_2}{k_{-1}} [S][Y]$$

so that

$$k_{obs} = \frac{k_1 k_2}{k_{-1}} = K k_2 \tag{2.6}$$

where K is the equilibrium constant between the intermediate SY and S and Y. The rate profile for this situation will be as represented in fig. 2.4(b). If K_0, $(k_0)_2$ and $(k_0)_{obs}$ refer to the unsubstituted derivatives S_0, then from (2.6)

$$\log \frac{k_{obs}}{(k_0)_{obs}} = \log \frac{K}{K_0} + \log \frac{k_2}{(k_0)_2} \tag{2.7}$$

Assuming the two steps are governed by Hammett laws:

$$\log \frac{K}{K_0} = \sigma \rho_1,$$

$$\log \frac{k_2}{(k_0)_2} = \sigma \rho_2,$$

then

$$\log \frac{k_{obs}}{(k_0)_{obs}} = \sigma(\rho_1 + \rho_2) \tag{2.8}$$

The overall reaction thus follows the Hammett equation, with the observed ρ value being the algebraic sum of the values describing the separate stages.

A modification of (2.2) may be that the decomposition of SY to give products involves intervention of a second reagent i.e.

$$SY + Z \xrightarrow{k_2} P$$

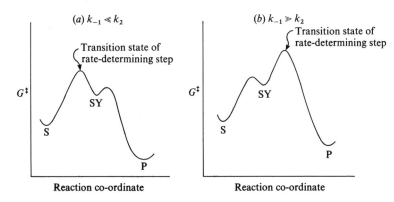

Fig. 2.4. Free energy profiles for reaction scheme (2.2).

The steady state concentration of SY now becomes

$$\frac{k_1[S][Y]}{k_{-1} + k_2[Z]}$$

so that

$$\frac{d[P]}{dt} = k_2[SY][Z] = \frac{k_1 k_2[S][Y][Z]}{k_{-1} + k_2[Z]}$$

In such a case, if $k_{-1} \ll k_2$, the rate is $k_{obs}[S][Y]$, which is independent of [Z], and equation (2.4) is applicable. Alternatively, if $k_{-1} \gg k_2$, the rate becomes dependent on [Z], being given by $k_{obs}[S][Y][Z]$ and equation (2.6) holds.

If the reaction scheme consisted of, not one as in (2.2), but several pre-equilibria followed by a slow irreversible step

$$\xrightleftharpoons{K_1} \xrightleftharpoons{K_2} \cdots \xrightleftharpoons{K_n} \xrightarrow{k_{n+1}}$$

then

$$k_{obs} = K_1 \, K_2 \, \cdots \, K_n \, k_{n+1} \tag{2.9}$$

and the observed ρ value is $(\rho_1 + \rho_2 \cdots \rho_n + \rho_{n+1})$.

Another case commonly encountered is that in which a substrate S yields a product P by two different paths.

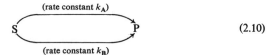

$$\text{(rate constant } k_A)$$
$$S \qquad\qquad P \tag{2.10}$$
$$\text{(rate constant } k_B)$$

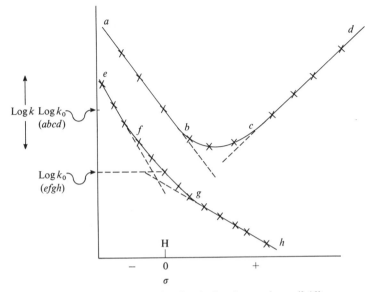

Fig. 2.5. Hammett plots for dual pathway scheme (2.10).

Here $k = k_A + k_B$ for the substituted compounds, and $k_0 = (k_A)_0 + (k_B)_0$ for the unsubstituted compound. If each step, considered separately, follows the Hammett equation, then, if k and k_0 are the observed rate constants for the overall reaction of the substituted compounds and the unsubstituted compound respectively,

$$\log \frac{k}{k_0} = \log \frac{k_A + k_B}{(k_A)_0 + (k_B)_0} \qquad (2.11)$$

When reaction A is sponsored by electron donors so that ρ_A is negative and reaction B by electron acceptors, so that ρ_B is positive, the resultant plot will be *abcd* shown in fig. 2.5.

There is no reason why the minimum rate should be k_0; we have arbitrarily chosen the unsubstituted compound as proceeding predominantly by reaction path A. Section *ab* represents the linear correlation of $\log k_A$ with σ; section *cd* that of k_B with σ. Section *bc* shows the changeover region from reaction A to B, where both mechanisms form a significant contribution to the overall reaction.

The second curve *efgh* illustrates the case where both ρ_A and ρ_B are negative.

The arguments presented in this section form the foundation of subsequent discussion relating experimental data to mechanisms by use of the Hammett equation. Equations (2.4) and (2.6) will be found of great importance; it is therefore particularly emphasised that their derivation should be understood and remembered.

2.6. Aromatic nucleophilic substitution. The bimolecular mechanism of aromatic nucleophilic displacement reactions is of the form [*2.2*]. The nucleophile is represented as negatively charged (for example, phenoxide), but it could equally well be a neutral amine, in which case the intermediate (**23**) will be a dipole from which a proton is lost.

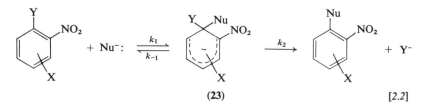

(**23**)   [2.2]

This has the same form as the general scheme (2.2). The presence of the nitro-group or similar strongly electron-withdrawing substituent in the *o*- or *p*-position is necessary to facilitate the stabilisation of the negative charge in the intermediate (**23**), which is often referred to as a Meisenheimer complex.

In the case of an active nucleophile Nu, and a good leaving group Y, the pattern becomes that of (2.4), because in such circumstances k_{-1} is much smaller than k_2; the observed rate constant is thus k_1. Most examples fall into this category, but instances are known where expulsion of Y becomes the rate-determining step as in (2.6).

An illustration of the former type of reaction is that of 1-chloro-4-nitrobenzene with an aliphatic amine RNH_2. Comparison of the canonical forms (**24**), (**25**) and (**26**) of the Meisenheimer complex, which we will take as an approximation to the transition state, with the canonical forms (**28**), (**29**) and (**30**) for resonance in *p*-nitroaniline, reveal similar electron delocalisations:

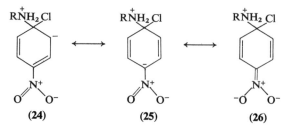

(**24**)   (**25**)   (**26**)

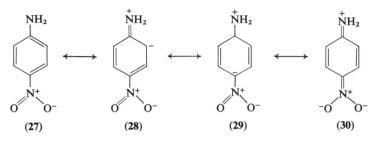

This leads to the prediction that the rates for such reactions will correlate with σ^- values, though we no longer have a side chain reaction; the benzene ring itself is involved directly, and its aromaticity is lost.

A suitable system to investigate the correlation has been provided by Greizerstein, Bonelli & Brieux (1962); these workers studied the rates of replacement of the chloro-substituent by piperidine in 4-X-1-chloro-2-nitrobenzene in benzene at 45 °C. The presence of the 2-NO_2 group is to activate the system to nucleophilic displacement; this will of course also impose a steric effect on the reaction centre, but its extent should be approximately constant throughout the series. Figure 2.6 shows the

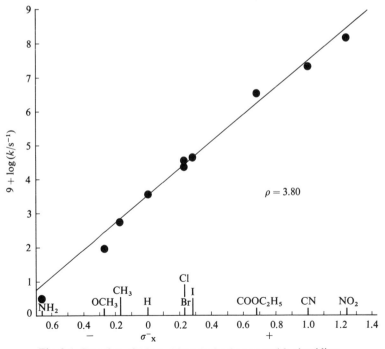

Fig. 2.6. Reaction of 4-X-1-chloro-2-nitrobenzene with piperidine.

correlation of the log second order rate constants with σ^-; the linearity is reasonable, particularly considering the wide range of reaction rates which have to be measured (1.49×10^{-1} and 3×10^{-9} l mol^{-1} s^{-1} for $X = NO_2$ and NH_2 respectively), which inevitably must introduce experimental errors. Indeed, the rates for NH_2 and OCH_3 are too slow to be measured at 45 °C; the values are obtained by extrapolation from higher temperatures. The large positive ρ value of 3.80 is in keeping with the production of extensive negative charge in conjugation with the p-substituents.

A variation on this mode of investigation is to study the effect of substituent changes in the nucleophile; clearly negative ρ values would be expected. Thus the reaction of m- and p-substituted benzylamines with 1-chloro-2,4-dinitrobenzene in ethanol at 45 °C (Fischer, Hickford, Scott & Vaughan, 1966) gives ρ -0.76, a low value because the benzene ring is insulated from the reaction site by a methylene unit (31); the consequence of which is that p-OCH$_3$ requires σ^0 for its correlation.

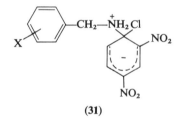

(31)

2.7. Aromatic electrophilic substitution. The accepted S_E2 mechanism for electrophilic attack on aromatic molecules fits into the generalised scheme (2.2); it is shown below for the reaction of electrophile E^+ with benzene.

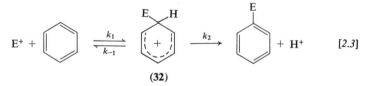

(32)

                                                             [2.3]

The loss of the proton is usually rapid, as demonstrated by hydrogen isotope studies; k_2 is thus much greater than k_{-1}, (2.4) holds, so that k_{obs} is k_1. (In the few instances where k_2 represents the rate-determining step, the aromatic substrate incorporates bulky groups adjacent to the position of reaction. The steric interaction energy involved in changing from the sp^3 hybridised Wheland intermediate (32), with E held out of the plane of the ring, to the sp^2 hybridised product matches the stabilisation energy of reattainment of aromaticity.)

In the previous section, it was found that nuclear reactivity of benzenoid compounds, in that case towards nucleophilic substitution, could be related to the Hammett equation established for side chain reactivities, involving use of σ^- constants. Extending this argument, it may be reasoned from comparison of the canonical forms making up the resonance hybrid of *p*-methoxyphenyldimethylcarbinyl carbonium ion, (34), (35) and (36) with those for the Wheland intermediate in *p*-nitration of methoxybenzene (37), (38) and (39), that σ^+ values will relate to rates of electrophilic substitution in the *m*- and *p*-positions of monosubstituted benzenes.

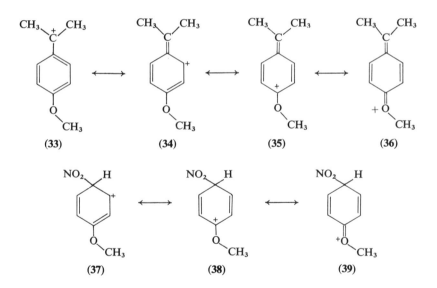

(33)   (34)   (35)   (36)

(37)   (38)   (39)

The appropriate equation should therefore be

$$\log \frac{k}{k_H} = \rho\sigma^+ \qquad (2.12)$$

where k is the rate constant for substitution at one nuclear position in a monosubstituted benzene, and k_H is the rate constant for reaction at one position in benzene itself. Thus k/k_H is a partial rate factor, f_p for a reaction at a p-position and f_m for a m-position, so that

$$\log f_m = \rho\sigma^+_m \qquad (2.13)$$

$$\log f_p = \rho\sigma^+_p \qquad (2.14)$$

Expressions (2.13) and (2.14) are particularly appropriate forms of the Hammett equation to use in this context, because most experiments relating to the rate of electrophilic attack on substituted benzenes have been carried out competitively, by comparing the relative amounts of products formed when two aromatic substrates together compete in the same reaction medium for a deficiency of electrophile. Indeed in some cases, such as nitration in organic solvents, this method is essential, because the rate of formation of the electrophile, rather than its rate of attack on the substrate, may be the slow step, complicating or nullifying the results of a kinetic investigation.

Table 2.5 gives data which show the validity of (2.13) and (2.14). Recorded here are log partial rate factors for various electrophilic substitution reactions (Stock & Brown, 1963) which are plotted against σ^+ values in fig. 2.7, giving acceptably straight line correlations, and very large negative values of ρ. The size of ρ is indicative of the extent of positive charge delocalisation in the aromatic nucleus; clearly, the more negative is ρ, the more the transition state of the reaction resembles the Wheland intermediate (**32**). For example, the transition state for ethylation is more 'reactant like' than that for bromination (fig. 2.8).

Figure 2.9 shows the correlation obtained for nitrations in sulphuric acid at 25 °C. For this medium, absolute reaction rates, rather than comparative ones, are available (Coombes, Crout, Hoggett, Moodie & Schofield (1970).

There are several marked deviations from the approximate straight line which are important. Firstly, m- and p-$\overset{+}{N}(CH_3)_3$ produce far slower reactions than their σ^+ values predict, but use of their σ values produces much better agreement. We have already commented that the discrepancy between σ^+ and σ values in this case is inexplicable; perhaps this latter result is an indication that experimental errors have arisen in the determination of hydrolysis data for m- and p-trimethylammonium phenyldimethylcarbinyl chlorides.

Secondly, at the opposite end of the scale, methoxybenzene also reacts many times slower than expected, in unison with other molecules anticipated to be particularly susceptible to nitration such as m-dimethoxybenzene, the xylenes, and mesitylene, and also probably fluorobenzene, biphenyl and toluene. This has been ascribed to the onset at this point of a diffusion-controlled reaction, in which the rate-determining step in this medium is the formation of a so-called 'encounter pair' (Coombes, Moodie & Schofield, 1968), rather than the Wheland σ complex.

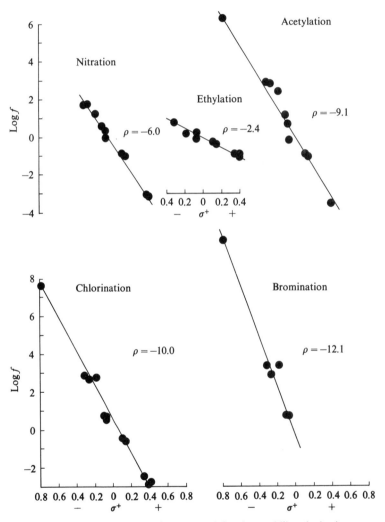

Fig. 2.7. Log partial rate factors *vs.* σ^+ for electrophilic substitution.

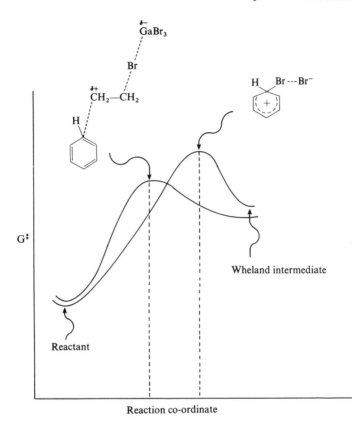

Fig. 2.8. Free energy profiles for ethylation and bromination of benzene.

Other complications arise for certain electrophilic substitutions. In table 2.6, data for the reactions of table 2.5 relating specifically to substitution in toluene are set out.

We can see that as the overall relative rate of toluene compared to benzene, $k_{toluene}/k_{benzene}$, for the first five reactions increases, so does the discrimination or *selectivity* between the *m*- and *p*-positions, p/m, also increase. (We cannot bring the *o*-position reactivity into the argument, because it will be affected by random steric interactions.) Nevertheless, the figures for the final two reactions clearly do not fit into this pattern; the values of $k_{toluene}/k_{benzene}$ are very small denoting that they are unselective electrophiles, but the amounts of *m*- and *p*-substitutions they produce in toluene are of the same order of magnitude as for nitration in

TABLE 2.5 *Electrophilic substitution reactions*

Reaction	Conditions	Log partial rate factors												
		p-OCH$_3$	p-CH$_3$	p-C(CH$_3$)$_3$	p-C$_6$H$_5$[a]	m-CH$_3$	m-C(CH$_3$)$_3$	p-F	p-Cl	p-Br	m-F	m-Cl	m-Br	ρ
1. Ethylation	C$_2$H$_5$Br, GaBr$_3$, C$_2$H$_4$Cl$_2$, 25 °C		0.78		0.35	0.19		−0.13	−0.23	−0.36	−0.94	−0.99	−1.06	−2.4
2. Nitration	CH$_3$COONO$_2$, (CH$_3$CO)$_2$O, 25 °C		1.69	1.76	1.58	0.32		−0.11	−0.89	−0.99		−3.08	−3.01	−6.0
3. Acetylation	CH$_3$COCl, AlCl$_3$ C$_2$H$_4$Cl$_2$, 25 °C	6.25	2.87	2.82	2.39	0.68	1.11	−0.18	−0.90	−1.08		−3.52		−9.1
4. Chlorination	CH$_3$COOH, 25 °C	7.67	2.91	2.60	2.78	0.69	0.73	0.64	−0.42	−0.54		−2.85	−2.73	−10.0
5. Bromination	CH$_3$COOH, H$_2$O, 25 °C	10.0	3.38	2.91	3.47	0.74	0.78							−12.1

[a] See § 4.6.

TABLE 2.6 *Stock and Brown's selectivity relationship for electrophilic substitution in toluene*

Reaction	$k_{toluene}/k_{benzene}$	% substitution			p/m	log f_p	log f_m	log f_p/f_m
		o	m	p				
1. Ethylation[a]	2.47	38.4	21.0	40.6	1.94	0.78	0.19	0.59
2. Nitration[a]	23	63.3	2.8	33.9	12.1	1.69	0.32	1.37
3. Acetylation[a]	128	1.17	1.25	97.6	78.1	2.87	0.68	2.19
4. Chlorination[a]	340	59.8	0.5	39.7	78.4	2.91	0.69	2.22
5. Bromination[a]	605	32.9	0.3	66.8	22.2	3.38	0.74	2.64
6. Bromination (Br$_2$, CH$_3$NO$_2$, FeCl$_3$, 25 °C)	7.1	71.1	1.6	27.3	17.0	1.07	−0.47	1.54
7. Nitration (NO$_2^+$BF$_4^-$, , 25 °C)	1.67	65.4	2.8	31.8	11.3	0.51	−0.85	1.36

[a] Condition as given in table 2.5.

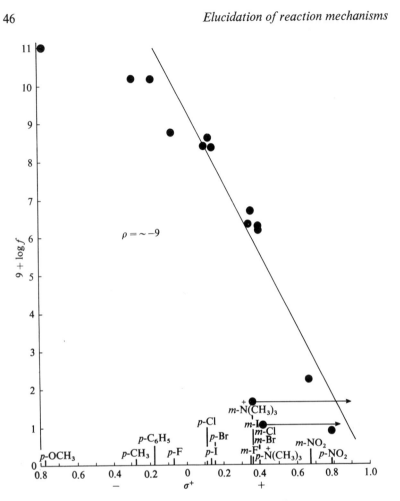

Fig. 2.9. Nitration in sulphuric acid medium.

acetic anhydride. The intramolecular selectivity is thus not commensurate with the intermolecular selectivity.

Let us demonstrate this discrepancy in a more quantitative manner. We can reason that a plot of $\log f_p$, a measure of intermolecular selectivity, against $\log f_p/f_m$, a corresponding measure of intramolecular selectivity, for a given substituent, in this case CH_3, in a series of electrophilic substitutions, should yield a straight line, slope m, since

$$\log f_p/f_m = \rho(\sigma^+{}_p - \sigma^+{}_m)$$
$$\log f_p = \rho\sigma^+{}_p$$

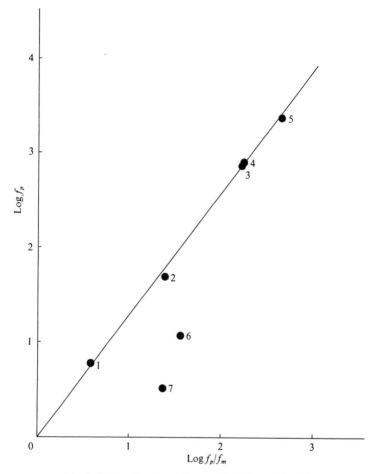

Fig. 2.10. The Stock and Brown selectivity relationship.

and thus

$$\log f_p = m \log f_p/f_m,$$

where

$$m = \frac{\sigma^+_p}{\sigma^+_p - \sigma^+_m} \qquad (2.15)$$

(2.15) is called the selectivity relationship (Stock & Brown, 1963); fig. 2.10 displays it from the figures in table 2.6.

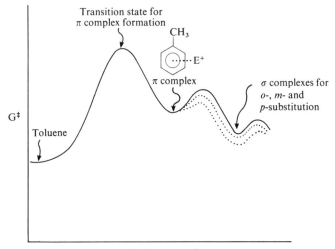

Fig. 2.11. Rate-controlling π complex formation for electrophilic substitution in toluene.

The slope defined by reactions 1–5 is 1.30, in agreement with 1.29 calculated from σ^+_m and σ^+_p using (2.15). However, the points for reactions 6 and 7 deviate markedly from the line. Olah (1971) has interpreted these results and others, particularly for certain types of alkylations as indicating that the rate-determining step for reactions 6 and 7 is the formation of π rather than σ complexes, so that the reaction profiles for *m*- and *p*-substitution appear as in fig. 2.11.

This seems a reasonable suggestion, because the stability of π complexes of benzene are known to be only minimally affected by the presence of methyl substituents attached to the benzene ring, which explains the low $k_{toluene}/k_{benzene}$ ratios. At the same time, formation of the products involves passage through σ complexes, giving rise to a preponderance of the more stable *p*-substituted toluene.

Contrary to this, other workers (Ridd, 1971) believe that these anomalies arise as a consequence of inefficient stirring of the reaction mixtures from which the relative rate $k_{toluene}/k_{benzene}$ is obtained; a description of the dispute which thus arose is beyond the scope of this book (see however problem 27).

The new conclusions, fundamental reinterpretations, and controversy currently concerning nitration and other electrophilic substitution reactions in aromatic molecules, which previously appeared so thoroughly documented and understood, particularly following the work of Hughes and Ingold (Ingold, 1953), cannot however go unremarked. There must certainly therefore be many surprises in store concerning reactions on which much less thought and effort has been expended.

2.8. Nucleophilic aliphatic substitution. Two distinct mechanisms exist for the displacement of a leaving group Z^- from its attachment to a saturated carbon atom in R by a nucleophile Y^- (Ingold, 1969).

$$RZ + Y^- \rightarrow RY + Z^-$$

Because the reacting carbon atom in R is saturated in the reactant and product molecules, the process is referred to as nucleophilic aliphatic substitution; obviously our Hammett investigations will involve junction of a substituted benzene nucleus. Notice also that the nucleophile may be a neutral molecule, of which water is a typical example.

The first mechanism, S_N1, consists of initial ionisation of RZ to yield the carbonium ion R^+ together with Z^-, and then rapid attack of Y^- to produce RY. This mechanism falls into the category (2.4) of the general scheme (2.2). The second, S_N2, involves synchronous rearward attack of Y^-, and cleavage of the C—Z bond. If the formation of the C—Y bond is in advance of cleavage of the C—Z bond, the carbon atom will become negatively charged. Conversely if C—Y bond formation falls behind C—Z bond breaking, the carbon atom will have carbonium ion character.

Since σ^+ constants have been designed from S_N1 hydrolysis of phenyl-dimethylcarbinyl chlorides, it is hardly surprising that in general S_N1 reactions are found to be correlated by such parameters. In hydrolyses where less stable carbonium ions are involved the reaction may proceed through an S_N1 mechanism if electron donor substituents are present, with a large negative ρ value, and through an S_N2 mechanism if electron acceptor substituents are present, with a smaller negative ρ value (fig. 2.12c). Such a scheme is expressed by the general equation (2.10) and a diagrammatic representation of it given as *efgh* in fig. 2.5. Pattern *abcd* would be obtained if the attacking nucleophile Y^- was sufficiently active so that bond making preceded bond breaking in the S_N2 process, with the result that the carbon atom undergoing attack bore a partial negative charge (fig. 2.12a).

An example of this curved type of plot (fig. 2.13) is afforded by the rates of solvolysis of substituted benzyl *p*-toluene sulphonates in 56%

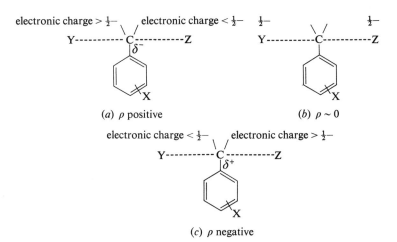

(a) ρ positive   (b) $\rho \sim 0$

(c) ρ negative

Fig. 2.12. Transition states in the S_N2 process $RZ + Y^- \rightarrow RY + Z^-$.

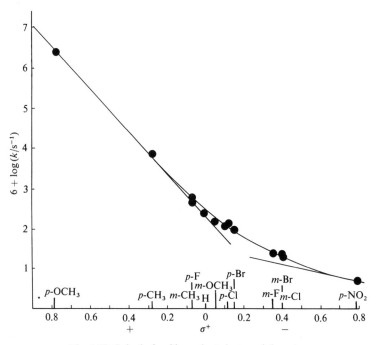

Fig. 2.13. Solvolysis of benzyl *p*-toluene sulphonates.

aqueous acetone at 25 °C (at constant ionic strength) (Fang, Kochi & Hammond, 1958).

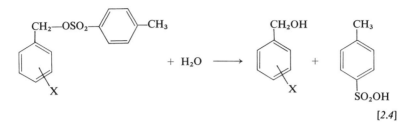

[2.4]

This falls into the *efgh* prototype of fig. 2.5, with the S_N2 transition state of the form of fig. 2.12*c*.

The authenticity of this type of approach has been demonstrated in elegant studies by Harris, Schadt, Schleyer & Lancelot (1969), on the rates of acetolysis of 2-arylethyl *p*-toluene sulphonates in acetic acid at 115 °C.

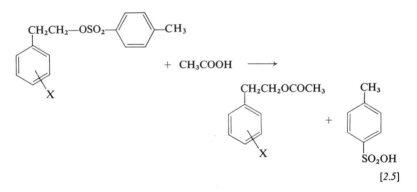

[2.5]

At first sight the reaction might be expected to yield a straight correlation with σ (σ^0 for *p*-OCH$_3$), of low ρ value due to the insulation of the reaction site from the ring by a methylene unit. In fact, the graph of log k_{obs} against σ shows pronounced curvature (fig. 2.14).

Clearly two mechanisms are operative ([2.6]): one is the anticipated S_N2 process with the correlation shown by the dotted line of slope -0.10, the other an S_N1 process proceeding by anchimeric assistance of the aryl group through the carbonium ion (**40**). No elimination occurs to yield olefinic byproducts, a common side reaction in many aliphatic nucleophilic substitutions.

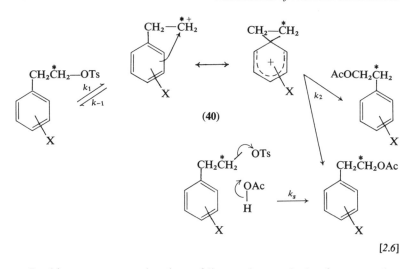

[2.6]

In this case, account is taken of 'internal return', that is to say, that
(40) is attacked, not only by acetic acid, but by tosylate anion, to
regenerate the starting material. The S_N1 scheme therefore falls into the
form of (2.6), so that

$$k_{obs} = (k_1/k_{-1})k_2 + k_S = Kk_2 + k_S \qquad (2.16)$$

k_{obs} may thence be divided in the following way. The dashed correlation
line in fig. 2.14 yields the value of $\log k_s$ for any substituent. Thus, for
$X = p\text{-OCH}_3$, $\log k_s$ is read off as -5.11; k_s is consequently 78×10^{-7}
s^{-1}, and hence, since k_{obs} is $4.00 \times 10^{-4}\,s^{-1}$, Kk_2 must be 3922×10^{-7} from
(2.16). In this way, the expression Kk_2 may be calculated for p-Cl, p-H,
$p\text{-CH}_3$ and $p\text{-OCH}_3$, the substituents corresponding to those compounds
for which an appreciable amount of reaction proceeds through the S_N1
route. A plot of $\log Kk_2$ against σ^+ for these substituents, also shown in
fig. 2.14, yields a good straight line of slope -2.4. This will be a composite
ρ, the algebraic sum of the ρ value for the $\log K$ correlation which will be
negative, and the ρ value for the $\log k_2$ correlation which will be positive.
There is evidence that k_{-1} is approximately independent of substituents,
because it describes reattachment of the group which ionised to its
original partner, both mutually enclosed in a solvent cage, and k_{-1} is
thus governed by statistical rather than reactivity factors, so that

$$\log Kk_2 = \log k_1 + \log k_2 + \text{constant}$$

For the overall ρ value to be negative, electron donor substituents

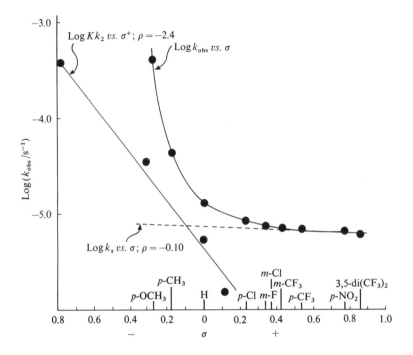

Fig. 2.14. S_N1 and S_N2 processes for acetolysis of β-arylethyl p-toluene sulphonates.

must accelerate the formation of the delocalised carbonium ion (**40**) more than they deccelerate the consequent attack of acetic acid on it.

The values of $(Kk_2/k_{obs}) \times 100$, the percentage reaction proceeding through the S_N1 route, are 18 for p-Cl, 42 for H, 82 for p-CH$_3$ and 98 for p-OCH$_3$. These figures agree well with those of 9, 38, 82, and 95, respectively, obtained by the use of ^{14}C scrambling data (if one of the carbons in the side chain, starred in [2.6], is labelled, its position in the final product will afford a key to the determination of the relative amounts of S_N1 and S_N2 reaction).

A further example of this partitioning of nucleophilic substitution into S_N1 and S_N2 routes is displayed in the work of Bosco, Forlani & Todesco (1970). It is well known that halogens in halogenobenzenes are very inert to nucleophilic displacement, unless strongly electron-withdrawing groups stand o or p to them. The introduction of an electron donor group would decrease reactivity even further in the benzene series; nevertheless, 1-halogeno-2-naphthols react readily with aniline producing 1-anilino-2-naphthol. The kinetics of the reaction for 1-bromo-2-naphthol in ethylene glycol at 100 °C with $NH_2C_6H_4X$ (X = *p*-OCH$_3$, *p*-CH$_3$, *m*-CH$_3$, H, *p*-Cl, *m*-Cl and *p*-NO$_2$) were found to be of the form

$$k_{obs}[\text{1-bromo-2-naphthol}] = k_1[\text{1-bromo-2-naphthol}] +$$
$$k_2[\text{1-bromo-2-naphthol}][\text{aniline}]$$

that is, the process has two components, one independent of the aniline, and the other dependent. For the latter, log k_2 was found to correlate with σ^- with a ρ value of -1.67. The reaction scheme proposed to explain these findings is given in [2.7].

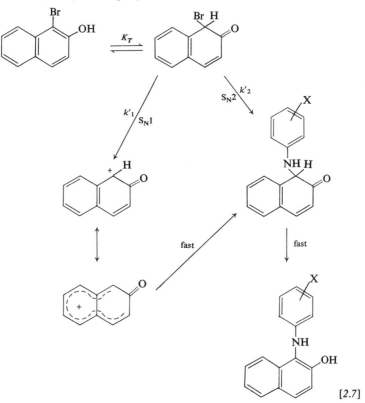

[2.7]

The difference in behaviour between 1-bromo-2-naphthol and 2-bromophenol thus originates from the former's ability to tautomerise to the ketone more readily than the latter, because while phenol loses all its aromatic stability in this process, naphthol retains that of one of its rings. The keto-tautomer can now undergo either an S_N1 reaction of rate constant k'_1, which is k_1/K_T from (2.6), or an S_N2 reaction of rate constant $k'_2 = k_2/K_T$ from (2.6), i.e.

$$\log k_2 = \log k'_2 + \text{constant}(\log K_T) \tag{2.17}$$

Aniline molecules are involved in this second process producing a negative ρ value in a σ^- correlation.

The final choice of example for this section comes from the work of Hart & Sedor (1967); in fact it combines an S_N1 process with electrophilic substitution. The reasoning employed by these workers here is intricate. The student may need to take some pains to follow it completely, but it repays careful study. The reaction is cyclodehydration of 2-phenyltri-arylcarbinols in 80% aqueous acetic acid containing 4% sulphuric acid at 25 °C.

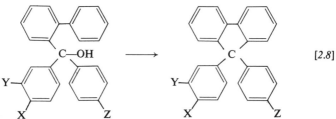

[2.8]

Figure 2.15 shows the plot of $\log k_{obs}$ against the sum of the σ^+ values of X, Y and Z. None of the compounds were in fact triply substituted, but were mono- or disubstituted in the positions indicated by X, Y and Z; the precise substitution of four of the compounds studied is indicated in fig. 2.15.

The pattern of this correlation is explained in terms of the reaction scheme below.

$$\text{ROH} + \text{H}^+ \underset{k_{-1}}{\overset{k_1}{\rightleftharpoons}} \overset{+}{\text{ROH}}_2 \quad \text{protonation}$$

$$\overset{+}{\text{ROH}}_2 \underset{k_{-2}}{\overset{k_2}{\rightleftharpoons}} \text{R}^+ + \text{H}_2\text{O} \quad S_N1$$

$$\left. \begin{array}{c} \\ \\ \end{array} \right\} \text{ROH} + \text{H}^+ \underset{k'_{-1}}{\overset{k'_1}{\rightleftharpoons}} \text{R}^+ + \text{H}_2\text{O}$$

$$\text{R}^+ \overset{k_3}{\longrightarrow} \quad \text{product} \quad S_E\text{Ar} \tag{2.18}$$

$$\text{R} \equiv$$

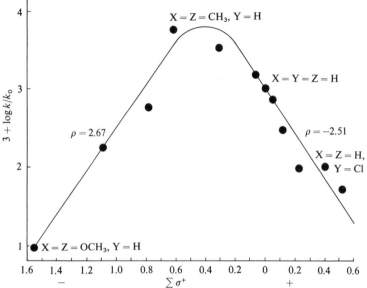

Fig. 2.15. Cyclodehydration of 2-phenyltriarylcarbinols.

The initial two equilibria, protonation and S_N1 carbonium ion formation, can be expressed in terms of a composite equilibrium constant k'_1/k'_{-1}. The final step of rate constant k_3 represents aromatic electrophilic substitution in which both the aromatic nucleus and the electrophile are supplied by the same molecule. Contrary to previous examples, however, substitution is varied in the electrophile, and the aromatic substrate remains unaltered.

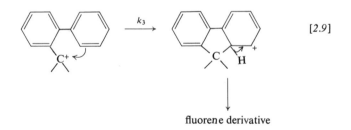

[2.9]

fluorene derivative

The correlation on the left-hand side of fig. 2.15 is consistent with this final step as the rate-determining one; it does not correspond to (2.6), however, because for this section of the correlation, the carbonium ions are more stable than the carbinols in the reaction medium. The effective reaction scheme is therefore

$$\text{ROH} \xrightarrow[\text{fast}]{\text{H}^+} \underset{\text{red}}{\text{R}^+} \xrightarrow{k_{\text{obs}}} \text{product} \tag{2.19}$$

and the ρ value is 2.67, positive as expected. This interpretation is substantiated by two experimental observations. Firstly, the red colour of the carbonium ion forms immediately and intensely on introduction of the carbinol into the medium; it then slowly fades as the colourless product is formed, and the rate of the reaction calculated from the rate of formation of product is identical to the rate evaluated by disappearance of this visible peak. Secondly, changing the medium to CH_3COOD–D_2O–D_2SO_4 does not alter the rate; since D_2SO_4 is a stronger acid than H_2SO_4, then clearly protonation cannot be implicated in the rate-limiting step.

As X, Y and Z are made less electron-donating, the carbonium ion becomes more unstable, its attack on the aromatic moiety gets increasingly urgent, and eventually we pass to the right hand correlation of fig. 2.15, where carbonium ion formation becomes rate-controlling. The general equation (2.4) now applies, giving us a second σ^+ correlation, this time with a negative ρ value of -2.51. The kinetic solutions for the implicated compounds are colourless showing that only transient carbonium ions are formed, while reaction in the deuterated solvent produces a threefold rate enhancement indicative of the involvement of acid in the rate-determining step.

2.9. Condensations of carbonyl compounds with amine derivatives. The polarisation of the π electrons in a carbonyl group towards the electronegative oxygen makes the carbon atom particularly susceptible to nucleophilic attack:

$$R\ddot{N}H_2$$

This initiating step is followed by a second, involving elimination of water from the intermediate carbinolamine. The overall reaction scheme is thus of the form

$$\text{>}C{=}O + RNH_2 \rightleftharpoons \text{>}C\underset{NHR}{\overset{OH}{\diagup}} \longrightarrow \text{>}C{=}NR + H_2O$$

[2.*10*]

The observed rate of such a reaction varies with pH, giving a characteristic 'bell-shape' rate–acidity profile. A typical profile is shown in fig. 2.16 for the formation of the Schiff base *N-p*-chlorobenzylidene aniline from *p*-chlorobenzaldehyde and aniline in aqueous solution at 25 °C. This shows a maximum rate at pH ~ 4, in the region of the pK_a of aniline. Similar maxima have been found for semicarbazone and oxime formation.

Jencks (1964) has done much work on the elucidation of this type of mechanistic scheme. The Hammett equation proved of considerable utility in his studies, and this prompts us to look into them in more detail, choosing for illustration the example of the reactions of semi-carbazide with substituted benzaldehydes. These were carried out in aqueous ethanol at 25 °C, using a large excess of semicarbazide so that pseudo first-order kinetics were obtained, with NaCl to maintain a constant ionic strength.

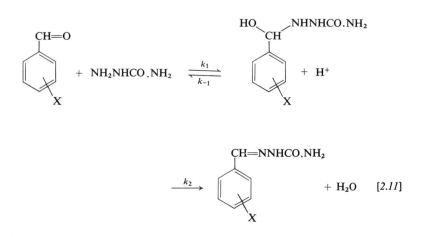

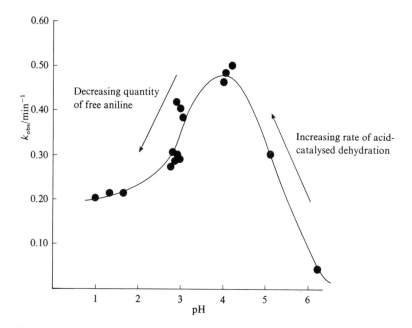

Fig. 2.16. The variation of the observed rate constant for formation of *N-p*-chloro-benzylidene aniline with pH.

When the solution is buffered to pH 8, the dehydration step is so slow that the equilibrium concentration of carbinolamine may be measured, and thus K discovered, where

$$K = \frac{k_1}{k_{-1}} = \frac{\left[\begin{array}{c}\diagdown \ \diagup OH \\ C \\ \diagup \ \diagdown NHR\end{array}\right]}{\left[\begin{array}{c}\diagdown \\ \diagup C{=}O\end{array}\right][NH_2R]} \qquad (2.20)$$

The correlation of log K with σ is shown in fig. 2.17; the ρ value is 1.81. It must be admitted that since there is through conjugation between the carbonyl group of benzaldehyde and a resonance donor substituent in the *p*-position, which is not present in the carbinolamine, a fit with σ^+ might have been expected. Reference is made to this point in table 3.8.

If acidity is now increased to between pH 6 and 7, the dehydration rate of the carbinolamine increases, and its concentration becomes vanishingly small, following the form of SY in the general equation (2.2);

nevertheless, the dehydration still remains the rate-limiting step as in (2.5) and (2.6). If we look at this step in detail, we find that it consists of a rapid protonation equilibrium, followed by elimination of water:

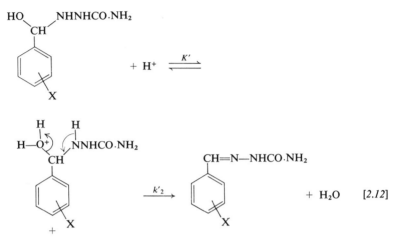

N-protonated form
(predominant)

for which $k_2 = K'k'_2$ (2.21)

Now the overall rate of reaction is evaluated as the rate of formation of semicarbazone, semicarbazide being present in large excess. Since there are no side reactions, this is equivalent to the rate of consumption of benzaldehyde:

$$d[{>}C{=}NHR]/dt = -d[{>}C{=}O]/dt = k_{obs}[{>}C{=}O]$$

So that in these low acidities where [2.12] is the rate-determining step:

$$k_{obs}\left[{>}C{=}O\right] = k_2\left[{>}C{\overset{OH}{\underset{NHR}{\diagup}}}\right][H^+]$$

and

$$k_2 = \frac{k_{obs}\left[{>}C{=}O\right]}{[H^+]\left[{>}C{\overset{OH}{\underset{NHR}{\diagup}}}\right]}$$ (2.22)

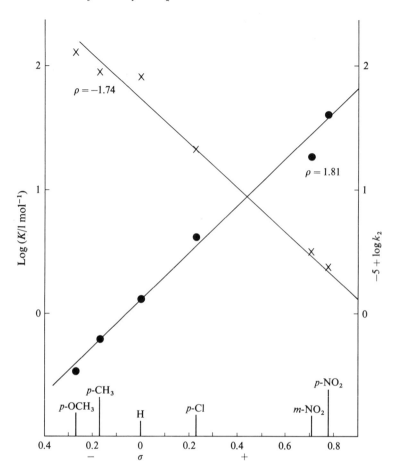

Fig. 2.17. Log K *vs.* σ for carbinolamine–carbonyl equilibria ●; log k_2 *vs.* σ for carbinolamine dehydration ×.

The quantity

$$\left[\underset{/}{\overset{\backslash}{C}}{=}O\right]\Big/\left[\underset{/}{\overset{\backslash}{C}}\underset{\backslash NHR}{\overset{OH}{\diagup}}\right]$$

can be calculated from K, which has been experimentally established: use of (2.22) therefore permits evaluation of k_2, the log values of which are also plotted in fig. 2.17. These correlate with σ values yielding a ρ value of -1.74. This suggests that substituents produce a large negative ρ value for the protonation equilibrium as expected, while loss of H_3O^+ is a largely synchronous process with ρ small.

Since

$$K = \left[\begin{array}{c} \diagup OH \\ C \\ \diagdown NHR \end{array} \right] \Big/ \left[\begin{array}{c} \diagdown \\ \diagup C{=}O \end{array} \right] [NH_2R],$$

we have from (2.22)

$$k_{obs}/[H^+] = k_2 \times K \times [NH_2R]$$
$$= k_2 K \times \text{constant} \qquad (2.23)$$

Therefore ρ for a plot of $\log k_{obs} = -1.74 + 1.81 \sim 0$ for these low acidities.

With increasing acidity, rate of dehydration also increases, until the pH equivalent to the pK_a of semicarbazide (3.7) is reached, producing a rate maximum. Whereupon the initial step of nucleophilic attack becomes

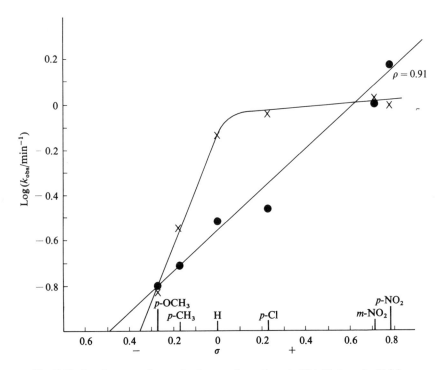

Fig. 2.18. Log k_{obs} *vs.* σ for semicarbazone formation at pH 1.75 ●, and pH 3.9 ×.

rate limiting as the concentration of nucleophile, and thus the rate of the reaction, diminishes. At these high acidities therefore

$$k_{obs} = k_1$$

and a graph of log k_{obs} against σ (fig. 2.18) has a slope of 0.91, indicating the expected acceleration by electron-donating substituents. Figure 2.18 also shows the curved plot obtained for log k_{obs} *vs.* σ at pH 3.9; here we observe a changeover in the rate-determining step – nucleophilic attack for benzaldehyde substituted with electron-donating groups, and dehydration with electron-withdrawing groups.

2.10. Free radical and multicentre reactions. All the reactions so far dealt with have involved heterolytic cleavage of bonds:

$$Z\!-\!Y \;\rightarrow\; Z^+ + Y^-.$$

Nevertheless, it is important to note that Hammett correlations may advantageously be applied to reactions involving homolytic cleavage:

$$Z\!-\!Y \;\rightarrow\; Z^{\cdot} + Y^{\cdot}$$

which suggests that even in this type of process, the transition state has considerable polar character, although it is true to say that ρ values in such cases tend to be of smaller magnitude than for heterolytic reactions. Some examples taken from Jaffé's review (1953) are given in table 2.7.

A number of reactions have also been studied in which bromine atoms or trichloromethyl radicals are responsible for α-hydrogen abstraction from substituted alkylbenzenes (Russell, 1958; Gleicher, 1968; Totherow & Gleicher, 1969).

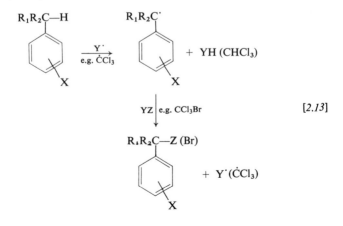

[2.13]

TABLE 2.7   *Free radical reactions*

Reaction	ρ
Decomposition of $XC_6H_4COOOC(CH_3)_3$ in $(C_6H_5)_2O$ at 100 °C	−0.90
Oxidation of XC_6H_4CHO with $C_6H_5COO\dot{}$ in $(CH_3CO)_2O$ at 30 °C	−0.49
Autoxidation of $XC_6H_4CHOHCOC_6H_4X$ at 10 °C	1.43

The best correlation is with σ^+ (fig. 2.19), revealing that the transition state involves carbonium ion character (**41**).

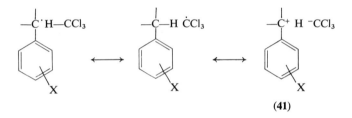

(41)

An allied type of reaction is the so-called multicentre or electrocyclic process in which, apparently, no charge separation is involved. We have already noted a σ correlation obtained with such a process (§ 1.6); an interesting variation of this scheme involves the attachment of the substituted nucleus to the alkyl group of the ester (Smith & Kelly, 1971).

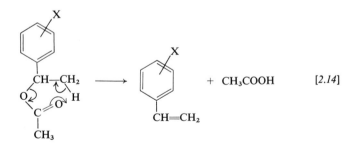

$$+ \ CH_3COOH \qquad\qquad [2.14]$$

An excellent correlation is found with σ^+ at temperatures of 327 °C in the vapour phase, ρ being −0.66. Again it is remarkable that agreement should be so good with constants defined from a reaction in a polar

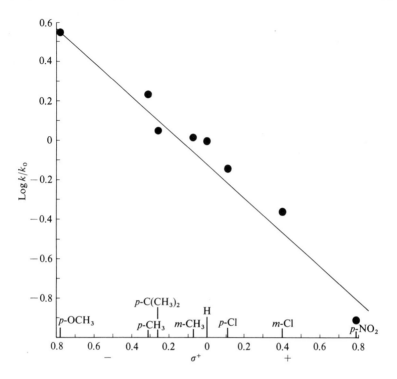

Fig. 2.19. Hydrogen abstraction by trichloromethyl radicals from neopentylbenzenes ($R_1 = H$, $R_2 = C(CH_3)_3$ in [*2.13*]).

solvent at room temperature. Clearly cleavage of the CO bond runs ahead of formation of the CC double bond, so that the transition state must involve a major contribution from (**42**).

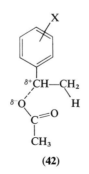

(**42**)

Analogous reasoning is supplied in the work of Bartlett (1970), who has investigated the effect of substituents on cyclobutane formation

in the reaction

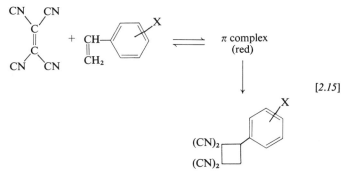

[2.*15*]

carried out in *o*-dichlorobenzene. The formation of the π complex is rapid and quantitative, and the overall rate may be followed by disappearance of the red coloration. Although this reaction proceeds very readily, synchronous electrocyclic thermal reactions to form cyclobutane derivatives usually give very poor yields (Woodward & Hoffmann, 1969). In this case, however, correlation of log k with σ^+, together with a large ρ value of -7.1, demonstrate that the bond redistribution is not synchronous, and the transition state is of zwitterionic form (**43**)

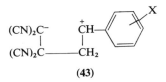

(**43**)

2.11. Conclusion. This chapter has contained a few important instances of the application of the Hammett equation to reaction mechanisms, selected from the many hundreds which exist in the chemical literature. Further reactions will be touched on later; ester hydrolysis and electrophilic additions to carbon–carbon double bonds, for example, which will be treated in some detail in chapter 3 in connection with the Taft equation. The following problem section deals with further reaction types – rearrangements, eliminations, solvolyses etc. – the elucidation of which depend precisely on the principles already expounded.

2.12 Problems

6. The pK_a value of 3,5-dimethyl-4-nitrophenol in water at 25 °C is 8.25; that of 3,5-dimethyl-4-cyanophenol is 8.21 (Wheland, Brownell & Mayo, 1948). Comment on these figures from the point of view of additivity of substituent effects, given that the pK_a of phenol is 9.99.

7. The effect of substituents on the log relative rates of acid cleavage, in aqueous methanolic perchloric acid at 51 °C, of phenyl trimethylsilanes is

p-N(CH₃)₂	7.5	m-CH₃	0.36	p-Cl	−0.87
p-OCH₃	3.18	H	0.00	p-Br	−1.00
p-CH₃	1.32				

(Eaborn, 1956)

Show that the reaction correlates with σ^+ and thus comment on the mechanism of the reaction.

8. The Lossen rearrangement of the potassium salts of acyl hydroxamic acids

$$R_1CO\bar{N}OCOR_2 \xrightarrow{OH^-,\ H_2O} R_1NCO + R_2COO^-$$

$$\downarrow \text{fast}$$

$$amine + CO_2$$

yields the following ρ values

$R_1 = XC_6H_4, R_2 = C_6H_5$:	$\rho = -2.59$
$R_1 = C_6H_5, R_2 = XC_6H_4$:	$\rho = +0.87$

(Renfrow & Hauser, 1937)

Show how the mechanism is consistent with these results.

9. The rates of acetolysis of *trans*-2-(*m*- and *p*-substituted benzoyloxy-)-cyclohexyl *p*-toluene sulphonates, in acetic acid with acetate ions present, yield a linear plot with σ with a ρ value of −1.00. If the carbonyl oxygen is labelled with ^{18}O, and the reaction product reduced with LiAlH₄, the *trans*-1,2-cyclohexane diol retains 50% of the activity (Gash & Yuen, 1969).

Comment on the mechanism of the reaction.

10. In a study of the hydrolysis of 4-substituted-2,6-dimethylbenzoyl chlorides in moist acetonitrile, a ρ of +1.20 for the reaction was obtained from a σ plot. When the reaction was conducted in the presence of perchloric acid, a correlation with σ^+ with ρ −3.90 was found (Bender, 1963). What mechanistic deductions can be made from these results?

11. The following table gives the reaction constants ρ, together with the isotope effects on the rate of the unsubstituted compounds ($X = H$) when deuterium is substituted for protium on the carbon atom adjacent to the ring, for the elimination reaction

Y	Relative rate	ρ	k_H/k_D
I	26 600	2.1	–
Br	4100	2.1	7.1
$OSO_2C_6H_4CH_3$	392	2.3	5.7
$\overset{+}{S}(CH_3)_2$	7.7	2.8	5.1
F	1	3.1	–
$\overset{+}{N}(CH_3)_3$	–	3.8	3.0

(Bunnett, 1969)

How do you interpret these results? What type of σ values would you expect to use for determination of ρ?

12. A Hammett plot for the second order rate constants of urethane formation from ethyl chloroformate and $XC_6H_4NH_2$ in anhydrous acetone has a ρ value of -5.56 for $X = p\text{-}OCH_3$, $p\text{-}CH_3$, $m\text{-}CH_3$ and H using σ values; a ρ value of -1.57 is found for $X = p\text{-}Br$, $m\text{-}Cl$, $m\text{-}NO_2$, $p\text{-}COOC_2H_5$ and $p\text{-}NO_2$ using σ^- where appropriate (Ostrogovich, Csunderlik & Bacaloglu, 1971).

Comment on the mechanism of the reaction.

13. Explain the results shown in the following table, which gives the ρ values for the acid-catalysed hydrolysis of esters under the conditions indicated:

Esters	Hydrolysis medium	ρ
(1) $XC_6H_4COOC_2H_5$	60% aq. acetone	+0.1
(2) $XC_6H_4COOCH_3$	99.9% sulphuric acid	-3.2 (using σ^+)
(3) $XC_6H_4COOCH(CH_3)_2$	99.9% sulphuric acid	+2.9
(4) $XC_6H_4COOC_2H_5$	99.9% sulphuric acid	curved σ plot with minimum at $\sigma = 0.6$
(5) $CH_3COOCH(C_6H_4X)CH_3$	30% aq. ethanol	-5.70 (using σ^+)

(Kershaw & Leisten, 1960; Hill, Gross, Stasiewicz & Manion, 1969)

3 The separation of inductive, resonance and steric effects; application of the Hammett equation to aliphatic systems

3.1. Introduction. The previous chapters illustrated how discrete sets of σ^+, σ and σ^- values afford, firstly, a sound basis for the general analysis of reaction mechanisms, and secondly, by the simple approximation that σ_p values reflect both resonance and inductive effects, and σ_m values inductive effects only, a method for estimation of such contributions to the electronic effects of individual substituents.

Such ideas are, however, open to criticism. Two important ones are set out below; the first has been mentioned before, and the second will probably have been anticipated by the reader: σ_m values must embody, if in reduced form, at least some influence from resonance as well as inductive effects; many reaction series might correlate with effective σ values intermediate between σ^+ and σ on the one hand, or σ^- and σ on the other.

But such considerations pose the question of whether attempts to incorporate further refinements really lead to a fundamentally better understanding of the equation's use and implications, and therefore justify the inevitable loss in simplicity. It must be appreciated that in general the introduction of additional parameters into an equation approximately defining some experimentally observed phenomenon cannot lead to a reduction in accuracy; the result must be an equation more exact than the original one. But such an increase in accuracy does not necessarily, therefore, indicate a valid theoretical improvement. Workers in the field appear to consider that the verification of this lies in rigorous statistical analysis of any suggested modifications of the Hammett equation. Such an analysis is required to discern a distinct and meaningful trend in experimental data which show at the same time a variation due to operator error and a variety of small order effects of unknown origin. A process of this kind involves precise detailing of definitions, the accumulation of as large a quantity of experimental

69

results as possible, and evaluation of them in terms of correlation coefficients, confidence limits, standard deviations and the like. This explains the rather indigestible look of advanced publications on the subject, and, because exponents often differ in their selection of data and their mode of assessment of them, the controversy and doubt which surround many of the conclusions drawn.

Let us take a simple but important example. In chapter 1 several pieces of evidence led us to the conclusion that I_p/I_m is approximately 0.95. Exner (1966) claims, however, that the ratio is 1.14, by considering the effect of substituents of the form CH_2X on the pK_a values of benzoic acids in 80% aqueous methylcellosolve and in 50% ethanol–water. But his deduction involves the assumption that the insulating methylene unit will be free of hyperconjugative effects, and leads to the unlikely conclusion that I_π (§ 1.4) is significant, and that groups like NO_2, CN, and SO_2CH_3 have zero resonance interaction with the benzene ring.

Thus, considering the overall evidence, the only inference which really seems valid is that the inductive effect as transmitted to the *m*- and *p*-positions is about equal in intensity, a deduction which is embodied in several theoretical approaches. But again, this is only an *approximate* conclusion; how much confidence can one have in the precision of other effects which are evaluated on the assumption that I_m is *accurately* equal to I_p in all cases?

The fact is that the systems to which the Hammett equation is applied are complex. General trends can be detected, and an approach to quantification made, but our present knowledge is completely incapable of coping with such systems with exactitude, because we cannot even recognise the origin and nature of many molecular interactions, let alone express them quantitatively. Considerable interpretative difficulty and consequently argument can arise when even the apparently simplest modification or extension of the σ–σ^+–σ^- pattern is attempted, however well justified theoretically, owing, at least in part, to the exaggeration of the inaccuracies in the Hammett laws when one seeks to perceive and elucidate a relatively small order effect usually only displayed by a limited number of substituents.

Keeping very carefully in mind these generalised conclusions, let us look at the way in which such investigations may be attempted.

3.2. The evaluation of inductive effects. The σ values which we have so far discussed contain contributions from both resonance and inductive effects. A very considerable amount of work, in which R. W. Taft (1956) has predominated, has been directed at the problem of dissecting σ values into contributions from these two components, and at quantifying steric contributions.

The first effort was initiated by Roberts & Moreland (1953) who synthesised a series of 4-substituted bicyclo[2.2.2]octane carboxylic acids (**44**) as models for the sigma framework of benzene, and measured their dissociation constants in 50% aqueous ethanol at 25 °C.

(**44**)

Table 3.1 lists the log K/K_0 values obtained from this system, the figures given being in fact those of Stock (Holtz & Stock, 1964; Baker, Parish & Stock, 1967) who extended and modified the original compilation. These values may now be plotted against σ_m (fig. 3.1); a reasonable correlation is obtained, lending validity to the approximations that σ_m reflects mainly substituent inductive effects, and I_π is negligibly small. The most substantial deviation is for OCH_3. This is not remarkable, for it has a large resonance effect, with a correspondingly important secondary contribution to σ_m; see (**5**), (**6**) and (**7**).

TABLE 3.1 σ_I *values from dissociation of* 4-X-*bicyclo* [2.2.2] *octane carboxylic acids*

X	log K/K_0	σ_I (log $(K/K_0)/1.65$)	σ_m	$(\sigma_m - \sigma_I)$	σ_p	σ_R $(\sigma_p - \sigma_I)$
H	0.00	0.00	0.00	0.00	0.00	0.00
OCH_3	0.47	0.28	0.12	−0.16	−0.27	−0.55
CH_3	−0.01	−0.01	−0.07	−0.06	−0.17	−0.16
C_2H_5	−0.02	−0.01	−0.07	−0.06	−0.15	−0.14
Cl	0.74	0.45	0.37	−0.08	0.23	−0.22
Br	0.74	0.45	0.39	−0.06	0.23	−0.22
$COOC_2H_5$	0.47	0.29	0.37	0.08	0.45	0.16
CN	0.93	0.56	0.56	0.00	0.66	0.10
NO_2	1.06	0.64	0.71	0.07	0.78	0.14
$\overset{+}{N}(CH_3)_3$	1.50	0.91	0.88	−0.03	0.82	−0.09

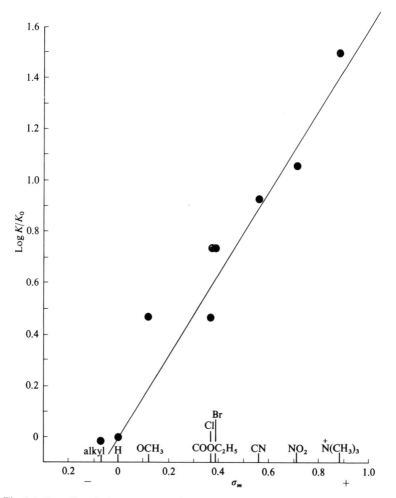

Fig. 3.1. Log dissociation constants of bicyclo[2.2.2]octane carboxylic acids *vs.* σ_m.

The slope of the graph (fig. 3.1) is 1.65. This is close to the ρ value for the correlation of benzoic acid pK_a's in this medium (table 1.4, [*1.5*]). σ_I may thus be defined as

$$\log K/K_0 = 1.65\,\sigma_I \qquad (3.1)$$

where σ_I *measures the contribution of the inductive effect to* σ_m *and* σ_p *values*, being considered, on the basis of previous arguments, to be the same for both (see § 1.4, together with 3.1).

The values obtained are given in table 3.1. It should be noted, because the nomenclature persists in the literature, that Roberts & Moreland originally designated the σ_I values obtained by this route as σ'. Expectedly, $(\sigma_m - \sigma_I)$ is usually small, being positive for $-R$ groups and negative for $+R$ groups.

The values of $(\sigma_p - \sigma_I)$, that is σ_R, confirm in somewhat more precise form the deductions set out in § 1.4; the strong $+R$ effect of OCH_3, Cl and Br, the relatively small $-R$ effects of $COOC_2H_5$, CN and NO_2, compared to their strong $-I$ effects, and the small but significant $+R$ effect of $\overset{+}{N}(CH_3)_3$.

What are surprising are the small σ_I values of CH_3 and C_2H_5, which are in fact indistinguishable from zero when experimental errors are taken into account. This is in line, however, with other results which suggest that the electronic effect of alkyl groups stems almost entirely from hyperconjugation which is only realised when they are attached to unsaturated carbon atoms (C^+ or $C{=}C$). Their inductive effect is thus small, and indeed in the case of CH_3 may even be in the opposite direction relative to hydrogen, so that the inductive order of electron release is $C(CH_3)_3 > CH(CH_3)_2 > CH_2CH_3 > H > CH_3$ (Jackman & Kelly, 1970).

Discussion of the inductive effect raises again the question of its mode of transmission from substituent to reaction site. Does it pass through the sigma bond framework or directly through space? The early experiments of Kirkwood & Westheimer (1938) demonstrated that the large differences between the first and second pK_a values of dibasic acids such as malonic or succinic could be ascribed to the latter form. In a further attempt to answer this question with regard to uncharged substituents, Wilcox & Leung (1968) have compared the dissociation constants of the bicyclic acids discussed above with those of the bicyclo[2.2.1]-heptane-1-carboxylic acids (**45**).

(**45**)

Systems (**44**) and (**45**) represent molecules in which the direct effect D is similar, but I_σ different, owing to the differing number of sigma bonds. The results of these and allied experiments are ambiguous because arbitrary decisions have to be taken regarding the fall-off factor for I_σ

as it passes from one bond to the next, and the effective dielectric constant within the molecule, to be utilised in the calculation of D. The conclusion seems to be that decided previously (§ 1.4), that both effects are operative, and apart from special cases, their complete separation and quantitative identification are impossible. One such special case, in which D can be clearly recognised, is in the pK_a values of ethano-bridged dihydro-anthracene carboxylic acids (46) (Golden & Stock, 1966).

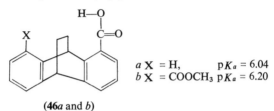

a X $= H$, $pK_a = 6.04$
b X $= COOCH_3$ $pK_a = 6.20$

(46a and b)

Here D must be large, because it represents the communication of the carboethoxy carbonyl group polarisation to the reaction site standing in close proximity to it, with resultant destabilisation of the anion:

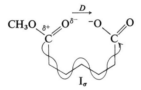

On the other hand, I_σ will be insignificant because many sigma bonds have to be traversed. The acid weakening produced by $COOCH_3$ clearly marks this angular dependence, reversing the normal influence of the group as found in the acidity of *m*- or *p*-carboethoxybenzoic acids.

Experiments where the distance between substituent and reaction site is short, but the number of σ-bonds separating the two is large, are relatively simple to design. Unfortunately, the converse situation, where the sigma bond separation is small, but distance through space is large, is obviously impossible to achieve. Nevertheless, an experiment revealing the role of I_σ has been carried out; it is described in § 4.9.

With regard to solvation influence on substituent effects it is note-worthy that Ritchie & Lewis (1962) have examined the correlation of the acidities of 4-substituted bicyclo-octane carboxylic acids (44) in a number of different solvents, and found good correlation with σ_I where the medium was acetone, methanol or ethanol, either neat or in aqueous mixtures. However, in dimethyl sulphoxide the correlation was poor, showing much scatter and suggesting unusual solvent substituent interactions.

The work initiated by Roberts & Moreland represents a very simple approach to the question of σ value dissection, but it fully illustrates the principles used by subsequent investigators. For example, the Taft equations, considered next, are derived from an extension of the Roberts–Moreland treatment in which the bicyclo-octane system X—C_8H_{12}—Y has shrunk to a single methylene group X—CH_2—Y, with the consequent introduction of steric as well as electronic interaction between X and Y.

3.3. Esterification and ester hydrolysis: the Taft equation. The $A_{Ac}2$ (bimolecular, acid-catalysed, acyl oxygen fission) and $B_{Ac}2$ (bimolecular, base-catalysed, acyl oxygen fission) mechanisms are predominant for acid- and base-catalysed ester hydrolysis under the majority of circumstances (Johnson, 1967):

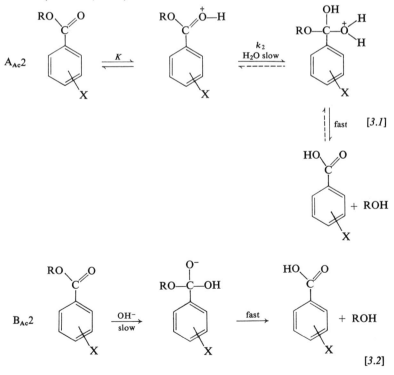

The ρ values for the $B_{Ac}2$ mechanism are appreciable. Table 1.4 gives 2.54 for 85% aqueous ethanol at 25 °C [*1.15*], and this is of the magnitude typically observed; in 60% aqueous acetone at 25 °C the value is 2.23. In contrast, the ρ values for the $A_{Ac}2$ mechanism are small; for example, 0.14 in 60% aqueous ethanol at 100 °C. Presumably, the reason for this is that the effect of X on the stability of the conjugate acid of the ester, relative to the ester itself, is compensated by its opposite influence on the extent of nucleophilic attack of water on this conjugate acid. Equation (3.2) expresses this in terms of the reasoning in § 2.5, equation (2.6):

$$\log k_{obs} = \underset{\substack{\rho_1 \text{ large} \\ \text{and negative}}}{\log K} \Bigg| \underset{\substack{\rho_2 \text{ large} \\ \text{and positive}}}{+ \log k_2}$$

$$\rho_{obs} = \rho_1 + \rho_2, \quad \text{where} \quad \rho_1 \simeq \rho_2 \qquad (3.2)$$

The difference in substituent effects on [*3.1*] and [*3.2*] has been utilised by Taft (1956) in a further attempt to measure inductive effects. There are a number of interesting and debatable points arising from this treatment, and it is worthwhile looking at it carefully, and in some detail.

In particular, it involves a second method of formulation of σ_I values for a wide variety of substituents, and investigation of their applicability to reactions of aliphatic compounds. That the Hammett equation does not apply to aliphatic compounds directly is clearly illustrated by fig. 3.2.

Here log rate constants for esterification of aliphatic acids (XCOOH) have been plotted against their pK_a values, and the result is a 'scattergram'. This lack of correlation may reasonably be assumed to be due to steric effects arising from the proximity of the substituents to the reaction site.

We have already seen that the rates of acid hydrolysis of *m*- and *p*-substituted benzoic esters are practically constant for a given medium, and thus are unaffected by changes in electronic factors. Therefore the wide variation of the rates for acid hydrolysis of aliphatic esters must be due to steric effects. A steric parameter E_s can hence be defined by equation (3.3)

$$\log \left(\frac{k}{k_0}\right)_A = E_s \qquad (3.3)$$

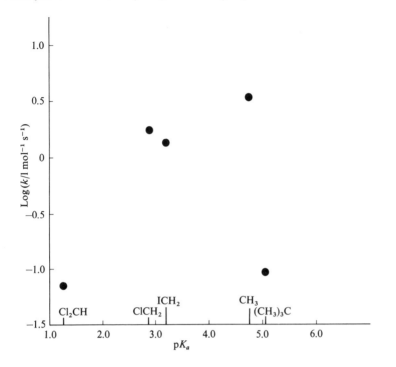

Fig. 3.2. Log rate constants for esterification of aliphatic acids, XCOOH, (C_2H_5OH, HCl, 14.5 °C) *vs.* pK_a values.

where k is the rate constant for acid hydrolysis of the substituted acetate XCH_2COOR, while k_0 is the rate constant for the unsubstituted ester CH_3COOR. Moreover, since reaction [*3.1*] is reversible, rates of esterification of the equivalent carboxylic acids, proceeding through the same transition state, may also be utilised. The E_s values are found to be generally independent of the nature of R and of the solvent in which the reaction is conducted. The values of E_s are negative, since substitution of H by a larger atom or group of atoms X will produce a decrease in rate due to steric interaction between X and the reaction site. In table 3.2 are collected some values of E_s for alkyl and halogen substituents.

These values are compared with ΔG^0_X, the conformational free energy of the group X in the monosubstituted cyclohexane $C_6H_{11}X$ (Eliel, 1965; Jensen, Bushweller & Beck, 1969). In the two equilibrating boat forms of cyclohexane, the substituent X may be either equatorial (**47**) or axial

T ABLE 3.2   *Steric parameters E_s and ΔG^0_X*

Hydrocarbon substituents (X in XCH$_2$COOR)

	E_s	$\Delta G^0_X/4.18$ kJ mol^{-1}
H	0.00	
CH$_3$	−0.07	−1.7
C$_2$H$_5$	−0.36	−1.8
CH(CH$_3$)$_2$	−0.93	−2.1
C(CH$_3$)$_3$	−1.74	> −4.4
C$_6$H$_{11}$	−0.98	−2.2
C$_6$H$_5$	−0.38	−3.1

Halogen substituents (X in XCH$_2$COOR)

X	E_s	$\Delta G^0_X/4.18$ kJ mol^{-1}	Van der Waals radii/nm	C—X bond length in CH$_3$—X/nm
F	−0.24	−0.28	0.14	
Cl	−0.24	−0.53	0.18	0.176
Br	−0.27	−0.48	0.20	0.191
I	−0.37	−0.47	0.22	0.210

(**48**), in the latter case encountering 1,3-diaxial hydrogen repulsions, which make the equatorial isomer the more stable one.

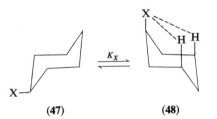

(**47**)   (**48**)

ΔG^0_X (= $RT \ln K_X$) thus represents the conformational free energy difference between the axial and equatorial isomers, and is a parameter which may be compared with E_s, in that both represent assessments of the bulk of groups. But examination of table 3.2, in which the ΔG^0_X values are also gathered, shows only a minimum of correspondence between the two quantities; in fact, about the only point of agreement is that C(CH$_3$)$_3$ is a very bulky group. This clearly emphasises that the steric effect contains contributions from several influences, whose relative allocations vary from system to system. Comparison of the data for the three halogens Cl, Br and I suggests that E_s is a function of the van der Waals' radius of X, a conclusion reached in general by Charton (1969). Alternatively, the decreasing steric interaction indicated by $\Delta G^0{}_X$ for these three halogens, indicates, by comparison with the bond lengths, that this quantity decreases with increasing distance by which a group is held from the cyclohexane ring, consequently reducing the steric interaction with the axial hydrogens.

In contrast to acid-catalysed hydrolysis, we remember that the base-catalysed hydrolysis is markedly dependent on electronic as well as steric factors. The transition states for the two reactions, $B_{Ac}2$ and $A_{Ac}2$, are seen from (**49**) and (**50**) to differ, formally at least, only in the number of protons involved, and this led Taft (1956) to suggest that steric effects, measured by E_s, should be very similar in both cases.

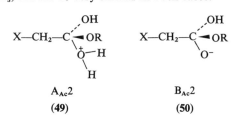

$$A_{Ac}2 \qquad\qquad\qquad B_{Ac}2$$
$$(49) \qquad\qquad\qquad (50)$$

Thus for basic hydrolysis of aliphatic esters, we can write

$$\log{(k/k_0)_B} = E_s + \sigma^* \rho \qquad (3.4)$$

assuming that steric effects can be treated independently of electronic effects, and that the latter can be accounted for in base-catalysed ester hydrolysis by a Hammett type law.

Combination of equations (3.3) and (3.4) leads to (3.5)

$$\{\log{(k/k_0)_B} - \log{(k/k_0)_A}\} = \sigma^* \rho \qquad (3.5)$$

The numerical value of the left-hand side of (3.5) is found to be similar, whether methyl, ethyl or menthyl esters are employed, or whether methanol or ethanol, neat or aqueous, is used as solvent. The agreement is not good, however, for purely aqueous medium.

The value of ρ is taken as 2.48, so that (3.5) becomes

$$\sigma^* = \frac{1}{2.48}\{\log{(k/k_0)_B} - \log{(k/k_0)_A}\} \qquad (3.6)$$

for the definition of σ^* values, a selection of which are given in table 3.3.

TABLE 3.3 σ^* *and* σ_I *values from aliphatic ester hydrolysis*

X (in XCH₂COOR)	σ^*	σ_I
H	0.00	0.00
OCH₃	0.52	0.23
CH₃	−0.10	−0.05
C₂H₅	−0.12	−0.05
C(CH₃)₃	−0.17	−0.08
F	1.10	0.50
Cl	1.05	0.47
Br	1.00	0.45
I	0.85	0.38
CN	1.30	0.60
COCH₃	0.60	0.27
N(CH₃)₃	1.90	0.90

The reason for the choice of 2.48 is usually given as 'putting the σ^* values on about the same scale as σ', since it is the average ρ value for alkaline hydrolysis of *m*- or *p*-XC₆H₄COOR, but as σ^* and σ therefore apply to different systems, such a statement seems rather vague. However, this is not important. Since the substituent X is insulated from resonance by a methylene group, and steric effects have been allowed for, σ^* values must measure inductive effects only, and should thereby permit the evaluation of σ_I values, and it is in this connection that their significance lies. This can be tested by plotting σ^* *vs.* σ_I, where the latter values are those measured by the Roberts & Moreland approach. Figure 3.3 reveals the correspondence to be excellent, the correlation being

$$0.45\,\sigma^* = \sigma_I \tag{3.7}$$

or

$$\sigma_I = \frac{1}{1.12}\{\log{(k/k_0)_B} - \log{(k/k_0)_A}\} \tag{3.8}$$

so that (3.7) or (3.8) can be utilised for calculation of further σ_I values for substituents of known σ^*, also shown in table 3.3.

It must be noted that certain approximations are involved in Taft's treatment for obtaining σ^* values. Firstly, the hydrogen atoms of the methylene unit, the latter ostensibly present for the specific purpose of removal of resonance interactions between X and the reaction site, may in fact hyperconjugate with the carbonyl group in the ground state of the ester to a variable extent depending on the nature of X, while in the

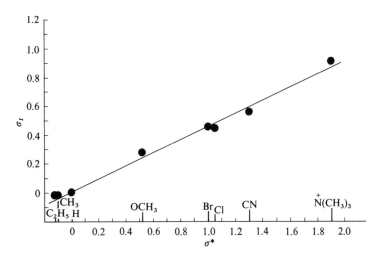

Fig. 3.3. Correlation of σ_I with σ^*.

transition state of base-catalysed ester hydrolysis, only hyperconjugation with X is present. **(51)**, **(52)** and **(53)** show this for X = CN.

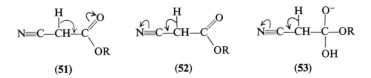

(51)   **(52)**   **(53)**

Indeed, Hancock (Hancock, Meyers & Yager, 1961), among others, has made an attempt to account for this type of interaction quantitatively.

Secondly, the assumption of equal steric effects in both acidic and basic media seems a drastic one, because of the possibility of solvation differences between the two different charge types **(49)** and **(50)**. Evidence has accumulated, for example, that the transition state in the acidic case involves a second water molecule (Yates & McClelland, 1967), being represented as in **(54)**.

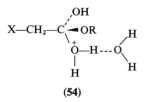

(54)

It is true that the medium in this case is water, a solvent for which correlations (3.3) and (3.6) do not afford good correspondence with values obtained in other media. Indeed, solvation differences between the transition states for acidic and basic reactions may be the source of this lack of correspondence, but it would be very strange if this difference in charge type did not lead to dissimilar solvation effects in all media.

Nevertheless, the practical validity of Taft's equation (3.6), in that it leads to values of σ_I which are compatible with those obtained by an independent route, justifies its use.

Further headway can be made. Taft found that log rate constants for a number of aliphatic reactions correlated linearly with σ^*. These included reactions as diverse as the acidities of XCH_2COOH and XCH_2OH, the catalysis of dehydration of acetaldehyde hydrate by XCH_2COOH, the catalysis of nitramide decomposition by XCH_2COO^-, and the acid-catalysed hydrolysis of

The implication is that here the steric interaction between substituents X and the reaction site is negligible, and since σ^* is proportional to σ_I, that resonance interactions are also absent.

Equations (3.3), (3.4) and (3.5) can also be applied to hydrolysis of esters of type $X_2CHCOOR$, $XYCHCOOR$ and X_3CCOOR. Table 3.4 shows some E_s and σ^* values for such groups.

TABLE 3.4 E_s and σ^* values for groups of type X_2CH, $XYCH$ and X_3C

X_2CH or $XYCH$ (in $X_2CHCOOR$ or $XYCHCOOR$)	X_3C (in X_3CCOOH)	E_s	σ^*
F_2CH		−0.67	2.05
Cl_2CH		−1.54	1.94
Br_2CH		−1.86	
	F_3C	−1.16	
	Cl_3C	−2.06	2.65
	Br_3C	−2.43	
$(CH_3)_2CH$		−0.47	−0.19
	$(CH_3)_3C$	−1.54	−0.30
$(C_6H_5)_2CH$		−1.76	0.41
$(CH_3)(t\text{-}C_4H_9)CH$		−3.33	
	$(C_2H_5)_3C$	−3.8	

TABLE 3.5 *Application of the Taft equation* (3.10) *to double bond addition reactions*

Reaction	ρ	δ
1. Chlorination, 25 °C	−6.67	0.95
	(Poutsma, 1965)	
2. Addition of 2,4-dinitrobenzenesulphenyl chloride		
((NO$_2$)$_2$C$_6$H$_3$SCl, acetic acid, 25 °C)	−2.84	0.66
	(Beverley & Hogg, 1971)	
3. Bromination, 25 °C		
(i) *trans*-substituted	−5.43	0.99
	(Dubois & Mouvier, 1965)	
(ii) *cis*-substituted	−5.43	1.48
	(Mouvier & Dubois, 1968)	
4. Addition of dichlorocarbene (chloroform, 0 °C)	−4.3	1.00
	(Skell & Cholod, 1969)	

A modification of (3.4) is

$$\log k/k_0 = \sigma^* \rho + \delta E_s \qquad (3.9)$$

which accounts for reactions in which steric effects cannot be neglected, but may be of a magnitude different from those in ester hydrolysis. Table 3.5 gives some examples of the application of (3.10) to reactions involving additions of electrophiles to substituted double bonds, for which it has been particularly useful.

3.4. The evaluation of resonance effects. A set of σ_I constants being established, their application to side chain reactivity data for *m*- and *p*-substituted benzenoid derivatives may be considered. In order to do this, (1.12) and (1.13) are rewritten as

$$\sigma_m = \sigma_I + \alpha\sigma_R \qquad (3.10)$$

$$\sigma_p = \sigma_I + \sigma_R \qquad (3.11)$$

$$\alpha = (\sigma_m - \sigma_I)/(\sigma_p - \sigma_I) \qquad (3.12)$$

introducing a factor α to describe the extent of communication of resonance effects to *m*-positions (Taft & Lewis, 1958).

Table 3.6 shows the complete dissection of σ_p values for a representative series of groups, utilising equations of the form of (3.11) for σ^0_p, σ^+_p and σ^-_p as well.

A crude theoretical calculation of α can be made. If transmission of an electronic influence by induction decreases by a factor of $\frac{1}{3}$ for every

TABLE 3.6   *Dissection of σ_p values into inductive and resonance contributions*

Substituent	Mechanism of electronic interaction	σ_p	σ^0_p	σ^+_p	σ^-_p	σ_I	σ_R	σ^0_R	σ^+_R	σ^-_R
NH₂	−I + R	−0.66	−0.38	−1.3[a]	−0.66	0.10	−0.76[a]	−0.48	−1.40[a]	−0.76
OCH₃	−I + R	−0.27	−0.12	−0.78[a]	−0.27	0.23	−0.50[a]	−0.35	−1.01[a]	−0.50
Cl	−I + R	0.23	0.27	0.11[a]	0.23	0.47	−0.24	−0.20	−0.36[a]	−0.24
CH₃	+I + R	−0.17	−0.15	−0.31	−0.17	−0.05	−0.12	−0.10	−0.26[a]	−0.12
COOC₂H₅	−I − R	0.45	0.46	0.48	0.68[a]	0.32	0.13	0.14	0.16	0.36[a]
NO₂	−I − R	0.78	0.82	0.79	1.27[a]	0.64	0.14	0.18	0.15	0.63[a]

[a] Enhanced or exalted σ values due to significant resonance interaction with reaction centre in ground or transition state.

σ bond through which it passes (Branch & Calvin, 1941) the ratio of the resonance effect at the m-position, conveyed there by induction from the adjacent o- and p-positions is $\frac{1}{3} + \frac{1}{3}$, i.e., 0.67.

In practice, however, this ratio is found to be much smaller, fortunately for our assumption that σ_m reflects predominantly the inductive effect of substituents. A value of 0.29 is considered appropriate for σ, 0.20 for σ^+ and σ^-, and 0.50 for σ^0 values. This variation from one reaction to another is of little significance. It stems entirely from the theoretical inadequacies and experimental errors which attend use of the Hammett equation, and make their presence felt when refinements of this type are attempted. For example, $(\sigma_m - \sigma_I)$ is a difference of two similar quantities, and is in consequence *enormously* susceptible to experimental inaccuracies and small differences in σ_I between one method of estimation and another. Moreover, it is a constant quantity for all scales. This is not strictly true for the σ^+ scale (see table 2.4) but deviations between σ_m and σ^+_m are small and random. Strong resonance donors show σ^0_p values less negative than σ_p, so for these α will be larger than 0.29, yielding an increase in the average value for all substituents. On the other hand, $(\sigma^+_p - \sigma_I)$ will be more negative than σ_p, producing a decrease in α. Similar reasoning shows that α for the σ^- scale will also decrease, this time in the case of resonance acceptors.

This is made clear by the following examples, the relevant figures being drawn from tables 3.1, 3.3 and 3.6:

$$OCH_3: \quad \sigma_m - \sigma_I = -0.16, \qquad \sigma_p - \sigma_I = -0.50, \qquad \alpha = 0.32$$
$$\sigma^0_p - \sigma_I = -0.35, \qquad \alpha^0 = 0.46$$
$$\sigma^+_p - \sigma_I = -1.01, \qquad \alpha^+ = 0.16$$
$$NO_2: \quad \sigma_m - \sigma_I = 0.07, \qquad \sigma_p - \sigma_I = 0.14, \qquad \alpha = 0.50$$
$$\sigma^-_p - \sigma_I = 0.63, \qquad \alpha^- = 0.11$$

Similar types of analyses, involving dissection into inductive and resonance components, have been attempted by Dewar & Grisdale (1962) and Swain & Lupton (1968). The former is described in some detail in the next chapter, because it involves a novel way for evaluation of the resonance component of substituents.

A corollary to the concept of partition of substituent effects into inductive and resonance contributions is that their differing modes of transmission through different side chains require different ρ values. Thus we may write

$$\log k/k_0 \text{ (or } \log K/K_0) = \rho_I \sigma_I + \rho_R \sigma_R \tag{3.13}$$

The right-hand side of equation (3.13) may be extended to

$$(\rho_D \sigma_D + \rho_S \sigma_S + \rho_R \sigma_R)$$

where the subscript D refers to the direct effect and S to I_σ. Indeed, no less than five distinct electronic effects with their associated modes of transmission have been suggested in this context (Dewar & Grisdale, 1962). However, the inherent simplicity of the Hammett equation asserts itself in that there is no evidence for assuming that ρ_D, ρ_S and ρ_R may be significantly different, and lend themselves to realistic separation one from another. Thus the equilibria and reactions in table 1.4 follow the Hammett equation with a high degree of precision, even though the reacting side chains registering the substituent effects are at a variety of distances from, and orientations with the benzene ring, and may be either in resonance with it or isolated from conjugative effects by one or more methylene groups. This conclusion is supported by a particularly noteworthy series of experiments on alkaline hydrolysis of compounds $XC_6H_4-C\equiv C-C\equiv C-C\equiv C-Si(C_2H_5)_3$ (Eaborn, Eastmond & Walton, 1971). Here, although the reaction site is held at a large distance from the ring by the rigid side chain, a correlation with $\sigma(\rho = 0.69)$ of high precision is obtained, using a wide variety of substituent types in the p-position.

The establishment of the general validity of comprehensive single value ρ constants is of major importance. Without this, as without the establishment of essentially solvent independent σ values (see the end of chapter 1), the Hammett treatment threatens to become so involved as to lose all practical utility.

3.5. The Yukawa–Tsuno equation. Values of σ^+ and σ^- are derived from standard reactions. When other reactions are considered, the σ_R components of the σ values relevant to their correlation might well display a variation from σ^+_R (and beyond) on the one hand, to σ^-_R (and beyond) on the other, each value expressing the extent to which the reaction centre demands or releases electrons by resonance.

Equations which would cater for this kind of situation were proposed by Yukawa & Tsuno (1959) and Yukawa, Tsuno & Sawada (1966) and investigated by them mainly with regard to σ^+ (3.14). Later, Humffray & Ryan (1969) examined the σ^- case (3.15)

$$\log k/k_0 \text{ (or } \log K/K_0) = \rho\{\sigma + r(\sigma^+ - \sigma)\} \qquad (3.14)$$

$$\log k/k_0 \text{ (or } \log K/K_0) = \rho\{\sigma + r(\sigma^- - \sigma)\} \qquad (3.15)$$

If r is zero, the equations simplify to (1.2) and (1.3). When r is unity, we have precise obedience to σ^+ or σ^-.

The use of these equations is best illustrated by considering a specific example, for which a good choice is the base-catalysed hydrolysis of triethylphenoxysilanes in 60% aqueous dioxan at 30 °C.

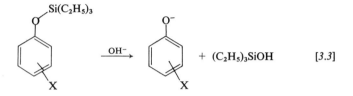

$$\qquad\qquad\qquad [3.3]$$

The transition state must be of the form (55), in which negative charge is accumulating on the oxygen of the substituted phenoxy group. Clearly the correlation will be of σ^- type and equation (3.15) is required. Figure 3.4 and table 3.7 illustrate the application of (3.15) to this system.

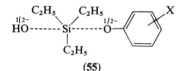

(55)

Initially, the straight line correlation, $\rho = 3.52$, is defined using those substituents for which σ^- is equivalent to σ, in this case $m\text{-}CH_3$, H,

TABLE 3.7 *Application of Yukawa–Tsuno equation to base-catalysed hydrolysis of* $XC_6H_4OSi(C_2H_5)_3$

X	$3 + \log (k/l\ mol^{-1}\ s^{-1})$	σ	σ^-	$(\sigma^- - \sigma)$	$\dfrac{\sigma +}{0.5\,(\sigma^- - \sigma)}$
H	0.423	0.00			
$m\text{-}CH_3$	0.330	−0.07			
$m\text{-}COCH_3$	1.646	0.38			
$m\text{-}CN$	2.553	0.56			
$m\text{-}NO_2$	2.755	0.71			
$p\text{-}N{=}NC_6H_5$	2.326	0.31	0.69	0.38	0.50
$p\text{-}COOC_2H_5$	2.443	0.45	0.68	0.23	0.57
$p\text{-}COC_6H_5$	2.911	0.43	0.88	0.45	0.66
$p\text{-}COCH_3$	2.714	0.50	0.84	0.34	0.67
$p\text{-}CN$	3.285	0.66	0.88	0.22	0.77
$p\text{-}CHO$	3.337	0.43	1.03	0.60	0.73

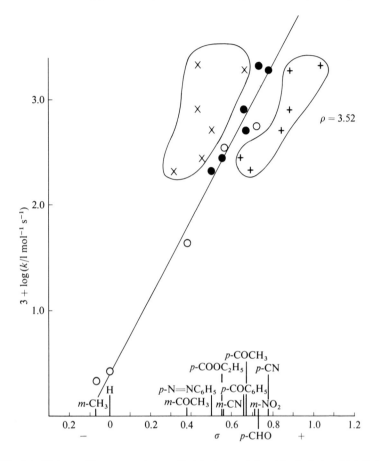

Fig. 3.4. Yukawa–Tsuno equation applied to base-catalysed hydrolysis of triethyl-phenoxysilanes. σ_m, $\circ$; σ_p, $\times$; $\sigma + 0.5(\sigma^-{}_p - \sigma_p)$, $\bullet$; $\sigma^-{}_p$, $+$.

m-COCH$_3$, m-CN and m-NO$_2$. Next, log rate constants for groups with unique σ and σ^- values, here p-N=NC$_6$H$_5$, p-COC$_6$H$_5$, p-COOC$_2$H$_5$, p-COCH$_3$, p-CN and p-CHO, are plotted using the two sets of values, both of which produce marked deviations from the correlation. Finally, the values of σ and σ^- for the individual substituents are introduced into (3.15), and the quantity r found which produces the best fit, in this case 0.50. This value is taken to indicate that marked but not complete breaking of the silicon–phenoxy oxygen bond has occurred in the transition state as shown (**55**).

TABLE 3.8 *Some typical r values for use in the Yukawa–Tsuno equation*

Reaction	ρ	r
Hydrolysis of $XC_6H_4C(CH_3)_2Cl$ (90% aq. acetone, 25 °C)	−4.54	1.00
Brominolysis of $XC_6H_4B(OH)_2$ (20% aq. acetic acid, 25 °C)	−3.84	2.29[a]
Methanolysis of $(XC_6H_4)_2CHCl$ (methanol, 25 °C)	−4.02	1.23
Bromination of XC_6H_5 (HOBr, $HClO_4$, 50% aq. dioxan, 25 °C)	−5.28	1.15
Nitration of XC_6H_5 (HNO_3, nitromethane or acetic acid, 25 °C)	−6.38	0.90
Ethanolysis of $(XC_6H_4)_3CCl$ (ethanolic ether, 0 °C)	−2.52	0.88
Basicity of $XC_6H_4N{=}NC_6H_5$ (20% aq. ethanol, 25 °C)	−2.29	0.85
Decomposition of $XC_6H_4COCHN_2$ (acetic acid, 40 °C)	−0.82	0.56
Beckmann rearrangement of $XC_6H_4C(CH_3)$: NOH (H_2SO_4, 51 °C)	−1.98	0.43
Semicarbazone formation of XC_6H_4CHO (25% aq. ethanol, 25 °C)	1.35	0.40[b]
Rearrangement of $XC_6H_4CH(OH)CH{=}CHCH_3$ to $XC_6H_4CH{=}CHCH(OH)CH_3$ (60% aq. dioxan, 30 °C)	−4.06	0.40
Decomposition of $(XC_6H_4)_2CN_2$ with benzoic acid (toluene, 25 °C)	−1.57	0.19

[a] This is the largest r value reported, but Norman & Taylor (1965) suggest that it is in serious error because there are large amounts of bromination at nuclear positions other than that from which the —B(OH)$_2$ group departs.

[b] See § 2.9 and problem 17.

This illustration demonstrates the use of the Yukawa–Tsuno equation under optimum conditions. A reasonable number of substituents have been examined for which σ and σ^- are equivalent, enabling accurate definition of ρ. Measurements have also been carried out on a further number of substituents for which $(\sigma^- - \sigma)$ is substantial, the lowest values being 0.23, and the others going as high as 0.60. The value of r is intermediate between unity and zero; the nearer r approaches either of these two latter values, the more uncertain is its separation from experimental error.

Unfortunately, the situation is rarely as clear-cut as this, particularly for reactions requiring use of (3.14). Table 3.8 shows some examples of the values of r obtained by the application of this equation, as selected, with one exception, by Ingold (1969).

Notice that a value of r greater than 1.00 implies greater conjugation of positive charge within the benzene ring than for the definitive system, which is the first entry. The most obvious point to be noted is that there is little correspondence between r and ρ, although a correlation would certainly have been expected, since both appear to indicate the extent of positive charge delocalised into the benzene ring in the transition state. It may be that r represents essentially a mathematical artifact with little theoretical significance. Certainly application of (3.14) can be made with much less certainty than (3.15), for a reason which is made plain by examination of the final columns of tables 2.2 and 2.4. For a σ^- correlation, a number of commonly occurring groups can be utilised, for which σ^- is very different from σ; this is not so in the case of σ^+ correlations. Here OCH_3 is the only simple group for which $(\sigma^+ - \sigma)$ is greater than 0.20. Amino substituents would be invaluable in these circumstances, but unfortunately the majority of reactions to which (3.14) is applicable are necessarily carried out in the presence of Brønsted or Lewis acids, so that N protonation or complexation would occur (see § 1.6), altering the character of this type of group to that of an electron acceptor.

The presence of acid in many of these reactions also suggests a further reason for the lack of correlation between r and ρ. Moreover, it provides, in the cases where r is intermediate between zero and unity, an alternative explanation for diversion of correlations from σ or σ^+ from that embodied in the Yukawa–Tsuno equation. Certain of the reactions will proceed via an initial free base-conjugate acid equilibrium, followed by a slow decomposition of the conjugate acid, obeying rate laws of the form of equations (2.6) and (2.8). Thus the decomposition of diazoacetophenones in acetic acid apparently follows scheme [*3.4*];

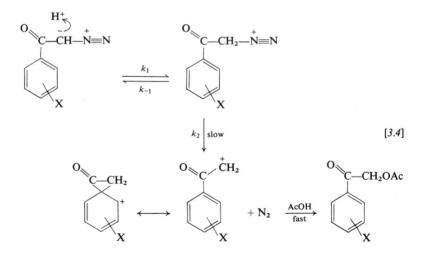

for which (2.8) should be modified to (3.16):

$$\log k/k_0 = \rho_1 \sigma + \rho_2 \sigma^+ \qquad (3.16)$$

ρ_1 refers to the pre-equilibrium, and ρ_2 to carbonium ion formation, both being positive. For substituents where σ approximates to σ^+, a good correlation will be found, but a deviation from correlation with both σ and σ^+ is to be expected for resonance donors in the *p*-position, the extent of which depends on the relative magnitude of ρ_1 and ρ_2.

Additionally, if in an aromatic electrophilic substitution the rate of decomposition of the Wheland intermediate enters the rate-determining expression [*3.5*], then the relevant rate equation is (2.6).

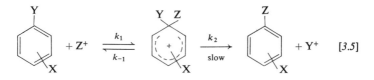

Although both the initial equilibrium and k_2 are governed by σ^+, the observed ρ value is again a composition of ρ_1 and ρ_2, where the sign of ρ_1 will be negative, but that of ρ_2 will be positive. Where Y is H the observed rate constant is generally k_1 ((2.4), also see discussion, § 2.7), but in other cases cleavage of the C—Y bond may well be slow. Thus protodesilylation [*3.6*] falls into the simple kinetic pattern of (2.4), but protodestannylation [*3.7*] and mercuridesilylation [*3.8*] have more uncertain rate dependencies and may be of the form of equation (2.6) (Norman & Taylor, 1965).

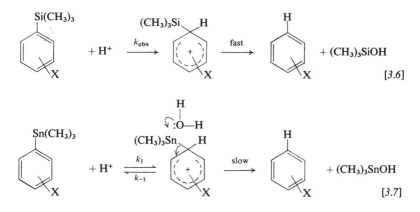

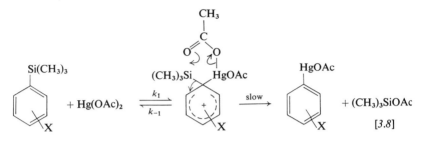

[3.8]

Consequently, if *r* is to be compared with ρ, only those values can be chosen which arise from reactions of similar mechanisms.

To conclude, it seems certain that while the Yukawa–Tsuno equation and related procedures, embodying variable resonance contributions from one reaction series to the next, will form the subject of much theoretical debate and enquiry, the simple approach using discrete σ^+, σ^- or σ sets will remain, as it is at present, the most practical basis for the interpretation of reaction mechanisms. Further data, directly relevant to the correlation of *r* with ρ, are discussed in the next chapter, §§ 4.5 and 4.6, where the extended selectivity principle is explained, and applied to substitution in the *p*-position of biphenyl.

3.6. Some attempts to evaluate *o*-substituent constants. Much work has been carried out on the elucidation of *o*-substituent constants, but not a great deal of progress has been made. This is due, certainly, to the difficulties of quantitative recognition of the contribution of steric factors and electrical effects, and, probably, to the independent variation of I_σ and D for *o*-substituents from one reaction to the next, in contrast to the apparently proportional variation of these effects for *m*- and *p*-substituents. Determination of σ_o values may either take the form of separate elucidation of steric and electrical effects, where both are present, or of examination of systems in which steric interactions are negligible.

The first approach may be made by Taft's method (1956). Hydrolysis data are available, not only for aliphatic esters, but also for *o*-substituted benzoates in both acidic and basic media, so that E_s and σ^* can be found by the use of (3.3) and (3.5), substituting for k_0 the rate constant for hydrolysis of the *o*-methylbenzoate ester. The value of 2.48 for ρ is appropriate here, for it is the average value for the basic hydrolysis of benzoates.

Table 3.9 shows the values thus obtained, the σ^* values being changed

TABLE 3.9 E_s and σ_o values

Substituent X	E_s	σ^*	σ_o	σ_p
H	–	–	0.00	0.00
OCH$_3$	0.99	−0.22	−0.39	−0.27
CH$_3$	0.00	0.00	−0.17	−0.17
F	0.49	0.41	0.24	0.06
Cl	0.18	0.37	0.20	0.23
Br	0.00	0.38	0.21	0.23
I	−0.20	0.38	0.21	0.27
NO$_2$	−0.75	0.97	0.80	0.78

to σ_o by the entirely arbitrary assumption that σ_o and σ_p are equivalent in the case of CH$_3$. Positive values of E_s mean, of course, that the o-substituents have less steric interaction with the reaction site than does CH$_3$.

While in general a good deal of confidence seems to be placed in the authenticity of Taft's analysis of aliphatic systems, there is much more scepticism with regard to its application to o-substituent effects, and to the validity of the E_s and σ_o values thus found (Charton, 1969). For example, the close correspondence between σ_o and σ_p for the more polar substituents, while accepted by some as evidence for the realism of the former, seems odd when one considers that, while the resonance effects from o and p might well be similar, the inductive effect, as transmitted either through bonds or space, must be quite different from the two positions.

A good illustration of the second mode of analysis is provided by the studies of Smith & Kelly (1971), who have reasoned that the steric effects of o-substituents might be considered small in the vapour-phase elimination reactions of esters and carbonates ([*1.21*] and [*2.14*]). In such cyclic eliminations the reaction site is compact and constrained away from steric interaction. The values of σ_o obtained from the rates of vapour-phase elimination of isopropyl benzoates, using a ρ value of 0.33 obtained from the correlation with m- and p-substituents, were: OCH$_3$, −0.53; CH$_3$, −0.16; H, 0; F, 0.16; Cl, 0.31; NO$_2$, 0.94.

The conclusions to be drawn from the measurements discussed in this section are obviously extremely tenuous, and there are almost as many significantly different values of σ_o for a given group as there are methods for defining them. Quite clearly, no scale of σ_o values can be constructed which will apply to a whole series of reaction types in the manner that σ_m and σ_p do.

3.7. Problems

14. Calculate E_s values from the following data (Taft, 1956) and thus comment on their degree of independence of the nature of the reaction, esterification or hydrolysis, and the solvent employed.

	Esterification (C_2H_5OH, HCl, 14.5 °C)					
Acid	CH_3COOH	CH_3CH_2COOH	$(CH_3)_3CCOOH$	$ClCH_2COOH$	$BrCH_2COOH$	ICH_2COOH
$k/l\ mol^{-1}\ sec^{-1}$	3.661	3.049	0.0909	2.432	1.994	1.713

	Hydrolysis (70 % aq. acetone, H^+, 25 °C)		
Ester	$CH_3COOC_2H_5$	$CH_3CH_2COOC_2H_5$	$(CH_3)_3C \cdot COOC_2H_5$
$10^5\ k/l\ mol^{-1}\ sec^{-1}$	4.47	3.70	0.128

15. The rate constants for alkaline hydrolysis of ethyl acetate, propionate and trimethylacetate in 70 % aqueous acetone at 25 °C are 4.65 × 10⁻², 2.20 × 10⁻² and 2.23 × 10⁻⁴ l mol⁻¹ s⁻¹ respectively. In 85 % ethanol at 25 °C the equivalent figures are 6.95 × 10⁻³, 3.55 × 10⁻³ and 2.54 × 10⁻⁵. Calculate the σ^* values for CH_2CH_3 and $C(CH_3)_3$.

16. The log rate constants (k/s⁻¹) for the alkaline hydrolysis of 4-X-1-methoxy-2-nitrobenzenes in water at 30 °C are:

X = H	−14.58	CF_3	−9.97
Cl	−13.37	NO_2	−6.62
Br	−13.40		

(Bowden & Cook, 1971)

Examine the application of the Yukawa–Tsuno equation to the reaction, and comment on the ρ value obtained.

17. The equilibrium constants for semicarbazone formation from substituted benzaldehydes in 25 % aqueous ethanol at 25 °C are as follows:

Substituent	p-OCH₃	p-CH₃	H	p-Cl	m-NO₂	p-NO₂
$K/l\ mol^{-1}$	0.34	0.62	1.32	4.14	18.3	40.1

With regard to the discussion of § 2.9, examine the applicability of the Yukawa–Tsuno equation.

18. The second order rate constants $k/l\ mol^{-1}\ s^{-1}$ for the addition of 4-X-2-nitrobenzenesulphenyl chlorides to cyclohexene in acetic acid at 25 °C are:

X = OCH₃	0.118	Cl	0.0371
CH₃	0.0484	CF₃	0.0124
H	0.0257	NO₂	0.0078

(Brown & Hogg, 1968)

By what mechanism is the addition proceeding?

The second order rate constants $k/l\ mol^{-1}\ s^{-1}$ for the addition of 2,4-dinitrobenzenesulphenyl chloride to 4- and 5-substituted cyclohexenes in glacial acetic acid at 30 °C

X X′

are as follows:

X	X′	$10^3\ k/l\ mol^{-1}\ s^{-1}$
H	H	11.6
C(CH₃)₃	H	10.4
CH₃	H	9.42
OCH₃	H	2.21
COOCH₃	H	1.30
Br	H	0.536
CH₃	CH₃ (*cis*)	4.52
COOCH₃	COOCH₃ (*cis*)	0.303

(Kwart & Miller, 1961)

Show that the reaction rates are influenced by the inductive effect of the substituents, and comment on the degree of quantitative correlation.

19. The following table gives the rates of hydrolysis of various methyl esters XCOOCH₃ in 77 and 96% aqueous sulphuric acid (Hopkinson, 1969). Relevant σ^* values are also included. Rationalise the results mechanistically.

X	σ^*	$10^3\ k/min^{-1}$ (77% H₂SO₄)	$10^3\ k/min^{-1}$ (96% H₂SO₄)
CH₃CH₂CH₂	−0.12	8.95	47.9
CH₃CH₂	−0.10	12.2	30.6
CH₃	0.00	23.3	12.1
C₆H₅CH₂	0.22	11.8	4.54
ClCH₂CH₂	0.38	6.65	0.358
ClCH₂	1.05	73.1	1.43
Cl₂CH	1.94	–	6.03

4 Application of the Hammett equation to data other than side chain reactivities of substituted benzenes

4.1. Heteroaromatic systems. Thus far, attention has been mainly restricted to consideration of reactivities of benzene derivatives, the only exception being a brief excursion into the realm of aliphatic chemistry by means of the Taft equation. Let us now see how the Hammett equation can be applied to reactivity data of heteroaromatic and polybenzenoid compounds and to measurements other than those of reactivity, particularly spectral determinations, on benzenoid systems.

For the situation in which a heteroatom is present in an aromatic ring, three possible approaches arise. This is illustrated using the example of the pyridine nucleus. The nitrogen atom may be taken as the reaction site, equivalent to a reacting side chain, and measurements made of the effect on its reactivity of substituents in the 3-(m)- and 4-(p)- positions.

Secondly, the nitrogen atom may be considered as a substituent, and its effect on electronic distribution in the ring determined by studying the reaction of a side chain m or p to it. Or, finally, both reacting side chain and substituent may be present in the pyridine nucleus, allowing the examination of additivity of effects.

All three approaches have been made, with what is generally considered to be a reasonable measure of success. This may be judged from the examples given below, chosen to illustrate the degree of correlation obtained.

4.2. The heteroatom as reaction site. Two obvious examples of reaction at pyridine nitrogen for which accurate data can be readily accumulated are the acid–base equilibrium and rate of quaternisation (Fischer, Galloway & Vaughan, 1964). In figs. 4.1 and 4.2 pK_a values in water at 25 °C and log k values for quaternisation with ethyl iodide in nitrobenzene at 60 °C, for a series of 3- and 4-substituted pyridines, are correlated with σ.

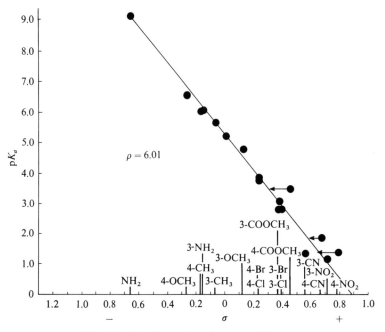

Fig. 4.1. pK_a values of substituted pyridines *vs.* σ.

For the protonation data the correlation is excellent. The high ρ value of 6.01 reflects the strong interaction between the substituents and the positive nitrogen atom within the ring itself; pyridine pK_a values are thus a very sensitive probe for substituent effects. The correlation for the quaternisation reaction is less good, but still reasonable. Here the ρ value is −2.94, of smaller magnitude, because in the transition state of the S_N2 reaction the positive charge is not fully developed on the nitrogen atom.

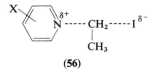

(56)

An additional point of some importance must be mentioned. The points for groups of potential (−*I* − *R*) type in the 4-position fall off the line in a direction which implies a reduction in their electron-withdrawing capacity as indicated by their σ values. Thus the apparent σ values from fig. 4.1 for 4-NO_2, 4-CN, and 4-$COOCH_3$ are 0.63, 0.58 and 0.28, rather

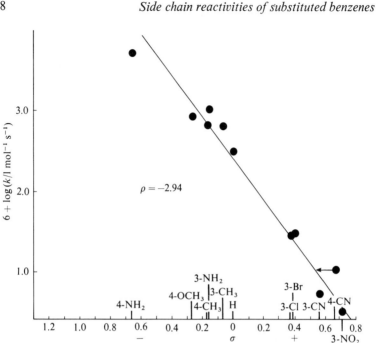

Fig. 4.2. Log k for rates of quaternisation of substituted pyridines *vs.* σ.

than 0.78, 0.66 and 0.45 respectively. The same sort of effect seems to occur in the quaternisation data, but here only the cyano group has been studied. These substituents appear to be exerting an inductive effect only, and hence the correlation is with σ_I values (compare the values given above with those for NO_2, CN and $COOCH_3$ in tables 3.1 and 3.3). This seems rational when it is considered that resonance withdrawal of electrons by such groups will be far more difficult from a nucleus containing an electronegative atom than from the benzene nucleus. But this argument leads to the deduction that the correlation with resonance donors should be with σ^+ rather than σ values, because in such cases there will be, apparently, greater resonance with the pyridinium ion than with the uncharged pyridine nucleus.

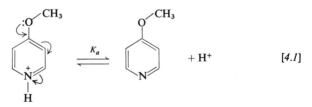

$$+ H^+ \qquad\qquad [4.1]$$

Thus for reasons which are not yet fully apparent, the pyridine system is a good deal less polarisable than benzene; that is, more reluctant either to donate or accept electronic charge by resonance.

4.3. The heteroatom as substituent. The nitrogen heteroatom is referred to as the aza substituent, while the protonated nitrogen is termed the azonium substituent. Appropriate σ values for the aza substituent can be derived, apparently quite simply, from dissociation constants of nicotinic and isonicotinic acids in water. In practice, however, the situation is complicated by tautomeric equilibria and zwitterion formation in such molecules. To appreciate this problem, and to demonstrate an approximate method of dealing with it, consider the example of isonicotinic acid (Blanch, 1966). The following scheme may be constructed to display the equilibria concerned:

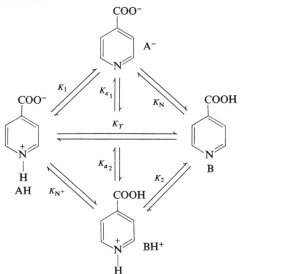

[4.2]

from which K_N and K_{N^+} are required for evaluation of σ_p for the aza and azonium substituents. However, direct experimental observation of the species involved as acidity is changed, leads only to pK_{a_1} and pK_{a_2} which are 4.86 and 1.84, respectively, in water at 22 °C.

$$K_N = \frac{[A^-][H^+]}{[B]} \tag{4.1}$$

$$K_{N^+} = \frac{[AH][H^+]}{[BH^+]} \tag{4.2}$$

Now

$$K_{a_2} = \frac{\{[AH] + [B]\}[H^+]}{[BH^+]} = K_2 + K_{N^+} \qquad (4.3)$$

For the determination of K_{N^+} a knowledge of K_2 is therefore required. For this a model system, in which the possibility of tautomerism is absent, is utilised. The one chosen is the protonation equilibrium of 4-carbomethoxypyridine.

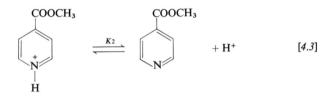

$$+ H^+ \qquad [4.3]$$

Writing K_2 for this equilibrium involves the assumption that the electronic effect of the $COOCH_3$ group is the same as that for $COOH$. Consideration of the σ_p constants for the two groups (both are 0.45) implies that this assumption is quite reasonable.

The measured pK_a value (pK_{a_2}) is 1.84, whence K_{a_2} is 0.0145, while the value of pK_2 from the model system [4.3] is 3.26. Note that this value, quoted by Blanch (1966) differs from that of Fischer, Galloway & Vaughan (1964) of 3.49, used previously. This difference is too large to be ascribed to the temperature difference of 3 °C between the two experiments, and is typical of the sort of discrepancy which often arises. It serves as an example of the relevance of the remarks made with regard to accuracy in § 1.2.

The value of K_2 is thus 5.5×10^{-4}, so that K_{N^+} and pK_{N^+} become 0.0139 and 1.86 respectively.

For calculation of K_N, the following relationship is used:

$$K_N = \frac{[A^-][H^+]}{[B]} = \frac{[BH^+]}{[B][H^+]} \times \frac{\{[AH] + [B]\}[H^+]}{[BH^+]} \times \frac{[A^-][H^+]}{[AH] + [B]}$$

whence

$$K_N = \frac{K_{a_1} K_{a_2}}{K_2} \qquad (4.4)$$

K_{a_1} is 1.38×10^{-5}, and from this value and the values of K_{a_2} and K_2 given above, K_N is found to be 3.63×10^{-4}, and thus pK_N is 3.44.

It is worthwhile noticing that

$$K_T = \frac{[B]}{[AH]} = \frac{K_2}{K_{N^+}} = \frac{1}{25} \qquad (4.5)$$

so that there is a preponderance of zwitterionic to neutral molecules in an aqueous solution of isonicotinic acid of 25:1. Knowledge of hetero-atomic substituent constants may thus be used to estimate tautomeric equilibria in the absence of relevant experimental data.

A similar reaction scheme for nicotinic acid yields K_N and K_{N^+} for the *m*-aza and *m*-azonium substituents, and use of the Hammett equation enables calculation of the appropriate σ values.

Table 4.1 shows the values thus obtained, together with values derived analogously from pyridine acetic acids in water at 25 °C, which are thus, strictly, σ^0 values. The agreement between the two sets of values is reasonable, which is to be expected, since σ^0 is equivalent to σ for resonance acceptors. Obviously, the azonium is a great deal more electron-withdrawing than the aza substituent; in fact the σ value of the former is by far the largest known, but surprisingly the difference $(\sigma_p - \sigma_m)$ reveals that they have the same electron-withdrawing capacity by resonance, and both attract the electrons predominantly by induction.

Measurements of rates of alkaline ester hydrolyses for a variety of substituted methyl benzoates, together with methyl pyridine 4-carboxy-lates (Campbell, Chooi, Deady & Shanks, 1970), reveal somewhat higher values for the aza substituent. These are also included in table 4.1. They

TABLE 4.1 *σ-values for the aza and azonium substituents*

Reaction	$\diagdown N \diagup$			$\diagdown \overset{+}{N} \diagup$ H		
	3	4	$(\sigma_p - \sigma_m)$	3	4	$(\sigma_p - \sigma_m)$
Ionisation of pyridine carboxylic acids (H_2O, 22 °C)	0.45	0.76	0.31	2.09	2.34	0.25
Ionisation of pyridine acetic acids (H_2O, 25 °C)	0.53	0.85	0.32	2.32	2.63	0.31
Alkaline hydrolysis of methyl pyridine carboxylates (85% aq. methanol, 25 °C)	0.65	0.96	0.31			

agree well with values obtained from ester hydrolyses in different solvents. Since they originated from direct measurement on systems in which the complicating effects of tautomerism are absent, they are probably the most accurate of the three sets, and will accordingly be used subsequently. A σ value of 0.75 was also calculated from these results, for the 2-aza substituent.

Deady (Deady, Foskey & Shanks, 1971) has applied these values in a remarkable correlation to the results of alkaline hydrolysis of diazine as

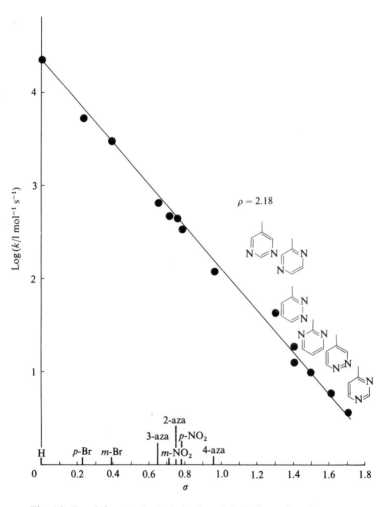

Fig. 4.3. Log k for ester hydrolysis of methyl diazine carboxylates *vs.* σ.

well as pyridine and benzene esters, carried out in 85 % aqueous methanol at 25 °C. The resultant plot, fig. 4.3, reveals, to a high degree of accuracy, the absence of steric interactions and accurate additivity of effects.

4.4. The heteroatom as part of the communicating system between reaction site and substituent. The only disposition of groups for examination of the transmitting properties of the pyridine nucleus, which avoids possible interactions between adjacent groups, is one in which the substituent and reaction side chain stand in positions 3 and 5. For such derivatives, the Hammett equation is followed quite accurately. Indeed, the final correlation given in the previous section revealed that steric interactions for aza substituents with side chain or other substituents are negligible. The ρ values obtained for such reactions are similar to those for the equivalent reactions in the benzene series, indicating that transmission of electronic effects through the benzene and pyridine systems is similar. For example, the results of Ueno & Imoto (1967) for the basic hydrolysis of 5-substituted ethyl nicotinates in aqueous ethanol reveal ρ to be 2.26, comparable to 2.54 (table 1.4) for the benzene series; the value of ρ for correlation of the pK_a values of the equivalent carboxylic acids in 50 % aqueous ethanol is 1.73, to be compared with 1.60 (table 1.4).

A moment's thought reveals that this is a necessary rather than a fortuitous correspondence, following from the additivity of substituent effects. Let us elucidate it in detail for the pyridine system bearing a reacting side chain, for which we may write:

$$\log k = \sigma \rho_N + \log k_N \qquad (4.6)$$

considering the pyridine nucleus as the transmitting system, k_N being the rate (or equilibrium) constant for the unsubstituted pyridine compound, and k that for the substituted pyridine nucleus.

However, the nucleus may be treated as benzene bearing an aza substituent, for which the additivity principle yields:

$$\log k = \rho(\sigma + \sigma_N) + \log k_0 \qquad (4.7)$$

where k_0 is the rate (or equilibrium) constant for the unsubstituted benzene nucleus bearing the same side chain. Combining (4.6) and (4.7),

$$\sigma \rho_N + \log k_N = \rho(\sigma + \sigma_N) + \log k_0 \qquad (4.8)$$

But

$$\log k_N = \sigma \rho_N + \log k_0$$

and hence

$$\rho = \rho_N \qquad (4.9)$$

This argument applies equally well to other systems, of course, particularly for five-membered ring heterocyclic compounds, as will be seen later. It is found to have the most interesting corollaries, when applied to the transmission of substituent effects through aromatic and heteroaromatic systems (§ 4.5), and to the Hammond postulate (§ 5.6). The protonation of pyridine 1-oxides is a further reaction of considerable interest. It may be considered in this section as an instance in which the reacting side chain is joined directly to the heteroatom of the transmitting system (Jaffé & Lloyd Jones, 1964).

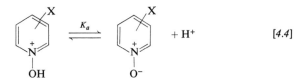

$$+ H^+ \qquad\qquad [4.4]$$

The system [*4.4*] is isoelectronic with the phenol–phenolate anion equilibrium (table 1.4), and therefore a correlation of pK_a's with σ^- is expected, and indeed found (fig. 4.4).

This yields a straight line of ρ value 2.10 (H_2O, 25 °C), with which the ρ value of 2.11 for the phenol equilibrium may be compared. Nevertheless, as the figure demonstrates, the correlation is poor for resonance donors in the *p*-position. The situation is somewhat improved by the employment of σ^+ values for such groups, a circumstance indicating enhanced resonance in the hydroxy-pyridinium ion.

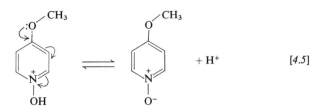

$$+ H^+ \qquad\qquad [4.5]$$

This is a logical result at first sight, but on reflection curiously at variance with the correlation of the pyridine pK_a values, where σ^+ was not required (§ 4.2).

4.5. Five-membered ring heteroaromatic compounds; the extended selectivity treatment. The electrophilic substitution of the five-membered heteroaromatic pyrrole, thiophene and furan has been extensively studied by Marino & co-workers (Clementi, Linda & Marino, 1970), for which

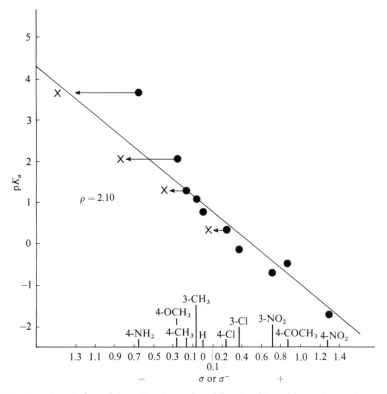

Fig. 4.4. Correlation of the pK_a values of pyridine 1-oxides with σ values. σ^+, $\times$; σ values (σ^- where appropriate) ●.

they have utilised the extended selectivity treatment of Stock & Brown (1963). In this procedure, equation (2.12) is employed, but the substituent is maintained constant, the log partial rate factors for that substituent in a given series of electrophilic substitutions being plotted against the equivalent ρ values. The point 0,0 should of course be on the line. Table 4.2 shows data for the 3(β)-position of thiophene, to which the extended selectivity treatment may be applied, as illustrated in fig. 4.5. The slope of the line, -0.52, is σ^+, which represents the electronic effect of substitution of S for CH=CH in the benzene ring.

The analogous treatment for the 2(α)-position of thiophene, which is a good deal more reactive than the β-position, and the α- and β- positions of furan also yields good straight line correlations, indicating the absence of steric interaction between reaction site and heteroatom. The σ^+ values

TABLE 4.2   *Electrophilic reactions at the β-position of thiophene*

Reaction	ρ	partial rate factor f_β
Bromination (CH_3COOH, H_2O, 25 °C)	−12.1	1.05×10^7
Chlorination (CH_3COOH, 25 °C)	−10.0	3.9×10^5
Acetylation (CH_3COCl, $ClCl_3$, $C_2H_4Cl_2$, 25 °C)	−9.1	1.35×10^4
Protodetritiation (CF_3COOH, 70 °C)	−8.2	5.8×10^3
Bromination {(Br^+), HOBr, $HClO_4$, H_2O, dioxan, 25 °C}	−6.2	1.6×10^3
Iododeboronation (I_2, H_2O, 25 °C)	−4.8	7.0×10^2
Protodesilylation ($HClO_4$, CH_3OH, H_2O, 50 °C)	−4.6	1.15×10^2

obtained are gathered in table 4.3, where the excellent correspondence with the σ^+ values obtained from the results of solvolysis of the appropriate 1-arylethyl acetates (problem 13) is also shown.

The heteroatoms are clearly very strong resonance donors in these five-membered ring systems, an effect which completely overrides their inductive withdrawal. The very reasonable linearity achieved by the extended selectivity treatment in these and related examples, when applied, for example, to electrophilic substitution at the *p*-position of methoxybenzene, thus argues strongly for a *constancy of resonance*

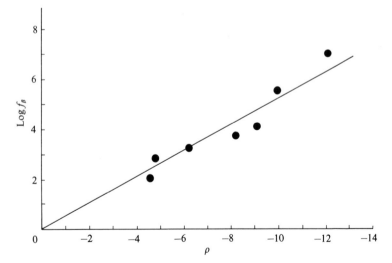

Fig. 4.5. The extended selectivity relationship for electrophilic reactions at the β-position of thiophene.

effect, whether the electron demand in the transition state, as measured by ρ, is large or small. It implies, in terms of the Yukawa–Tsuno equation, that variation of r in equation (3.14) from unity is largely due to experimental error, and thus provides strong support for the conclusion expressed in the final paragraph in § 3.5.

TABLE 4.3 σ^+ values for thiophene and furan

		From electrophilic reactivity	From solvolysis of 1-arylethyl acetates
Thiophene	σ^+_α	−0.79	−0.84
	σ^+_β	−0.52	−0.47
Furan	σ^+_α	−0.93	−0.94
	σ^+_β	−0.44	−0.49

Treating these rings as transmitting systems, good correlations have been reported for both side chain and electrophilic substitution reactions at the 2-position of thiophene with substituents in the 4- and 5-position, for which σ_m and σ_p values are utilised respectively (Butler, 1970). Significantly, the ρ values obtained are similar to those for the equivalent reactions in substituted benzenes; the reason for this is a further example of additivity of substituent effects accountable using the procedure illustrated in equations (4.7), (4.8) and (4.9).

Reflections on this theme produce a significant deduction. It is inappropriate and a tautologous argument to accept the tenets of the Hammett equation, and then to use it to compare the transmission of substituent effects through different systems. This is made clear by consideration of the application of the Hammett equation to the ionisation of pyrrole 2-carboxylic acids in water at 25 °C (Fringuelli, Marino & Savelli, 1969), where the above procedure produces a linear correlation, but of slope 1.65 rather than 1.00. The pK_a of the unsubstituted acid is 4.50, that of the 5-NO$_2$ is 3.22 and of the 5-CH$_3$ is 4.88. Substituting these figures into the appropriate form of equation (4.7), using the pK_a value of benzoic acid as 4.21, yields values of σ, for the replacement of NH for CH=CH in benzene, of −0.29, +0.21 and −0.50 respectively, a variation for which the explanation is not immediately apparent, and completely at variance with the Hammett equation assumption of a constant substituent effect.

4.6. Polycyclic systems: biphenyl. The application of the Hammett equation to side chain reactivities of polybenzenoid systems clearly represents an increase in complexity over the equivalent benzene systems, which is due to the greater choice of substituent and side chain positions.

Relevant data are limited in the literature, but some are available, mainly for the two simplest analogues, the biphenyl and naphthalene systems. Biphenyl itself can be treated very simply by considering it as a phenyl (C_6H_5) substituent attached to the benzene ring, and σ_m and σ_p values ascribed to it in the usual way (see table 2.4 and problem 2).

Electrophilic substitution of biphenyl displays an interesting pattern, worthy of detailed comment. In the last chapter, it was indicated that application of the Yukawa–Tsuno equation to σ^+ correlations (3.14) was ambiguous, and that however desirable a sliding scale of σ^+ values would be theoretically, a discrete set of such values was generally quite adequate for the description of most reactions involving electron-deficient transition states. This conclusion was given strong support in the previous section by application of the extended selectivity relationship to electron-donating systems. One is left however with the disquieting feeling that employment of this treatment with a resonance donor in the p-position for a series of such reactions with essentially similar mechanisms, for which the measured rate constants were free of large experimental error, *should* produce curvature implying a variable σ^+ value for the resonance donor under consideration. If under these optimum conditions, a linear relationship were still observed, then the correct conclusion to be drawn might well be that equations of Hammett type were insensitive to such variable effects. That this conclusion is untenable, and that the treatment is apparently quite capable of discerning a systematic variation where one exists, is shown by consideration of the extended selectivity relationship for electrophilic substitution in biphenyl (Stock & Brown, 1963), which from fig. 4.6 is shown to produce a curve. The partial rate factor for bromination ($\rho = -12.1$) in the p-position is 2920 giving an effective σ^+ value of -0.29, for which the r value in the Yukawa–Tsuno equation (3.14) is 1.6, while for ethylation ($\rho = -2.4$), the partial rate factor of 2.23 yields a σ^+ value of -0.14, and an r value of 0.8. The dotted line is that defined by a non-variable σ^+ value of -0.18.

This variable resonance effect comes from twisting of one of the phenyl rings out of the plane of the other; in the reactions yielding a ρ value of least negativity, the angle of twist will be large, but as ρ becomes increasingly more negative, denoting increasing positive charge on the

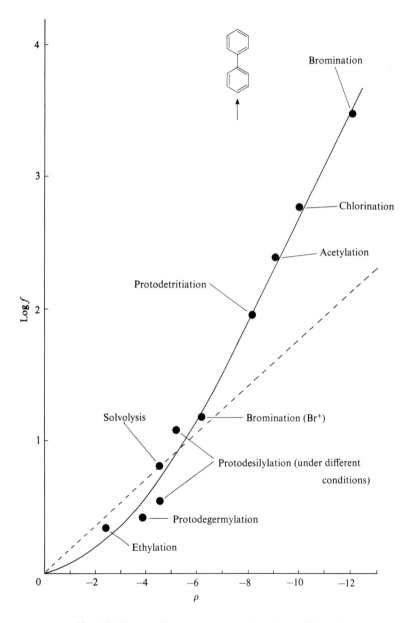

Fig. 4.6. The extended selectivity relationship for biphenyl.

ring in the transition state, this ring and the phenyl substituent assume a planar conformation with a resultant increase in stabilising conjugative interaction. Figure 4.7 shows the corresponding graph for fluorene which may be considered geometrically as a biphenyl in which the two rings are forced into coplanarity by the linking CH_2 group. The linearity of this correlation thus confirms the interpretation of the biphenyl example.

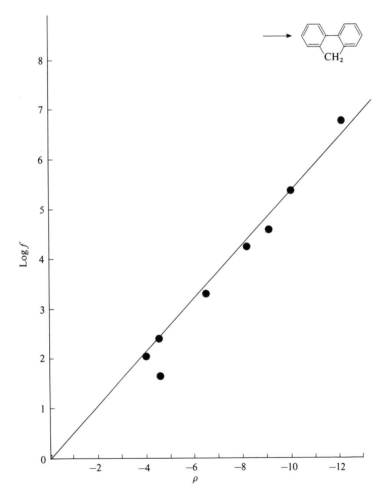

Fig. 4.7. The extended selectivity relationship for fluorene ($\sigma^+ = -0.54$).

This result is of considerable significance. The proportionality between ρ and r in the biphenyl case appears to confirm that the lack of such a proportionality in the other examples of σ^+ correlations shows that in these cases r is mainly indicative of experimental error alone, or due to the stepwise nature of the reaction as described in § 3.5.

4.7. Polycyclic systems: naphthalene. Naphthalene consists of a benzene ring with a 'benzo' substituent. It is found that the effect of substituents in the 3- and 4-position on the reactivity of side chains in the 1-position are correlated excellently by σ_m and σ_p values, respectively, as derived from the benzene system. This is illustrated in fig. 4.8 for the alkaline hydrolysis of ethyl 1-naphthoates (**57**) in 85 % aqueous ethanol at 50 °C (Fischer, Mitchell, Ogilvie, Packer, Packer & Vaughan,)58).

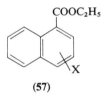

(**57**)

The ρ value is 2.21, to be compared with 2.54 for the equivalent process in the benzene system; we conclude from this that the transmission of effects between the 1- and 3- and 4-positions in benzene on one hand and naphthalene on the other is similar, yet another example of the validity of the type of treatment exemplified in equations (4.6) to (4.9). The corollary is that interaction between a 4-substituent and the 5-H in (**57**) is minimal for all cases studied.

Clearly, this approach is limited in its applicability. How can we explain the effect of substituents in the 5, 6, 7 or 8-positions of naphthalene on reactivity of a side chain in the 1 or 2-position? To answer this question, it is necessary to discuss very simple, 'back-of-an-envelope' calculations which yield surprisingly accurate estimates of the relative reactivities of conjugated hydrocarbons. These can be used in interpretation of the results of application of the Hammett equation to substituted derivatives of such systems.

4.8. Non-bonding molecular orbital theory. The NBMO theory has been developed from Hückel theory using the PMO (perturbational molecular orbital) method (Dewar, 1969).

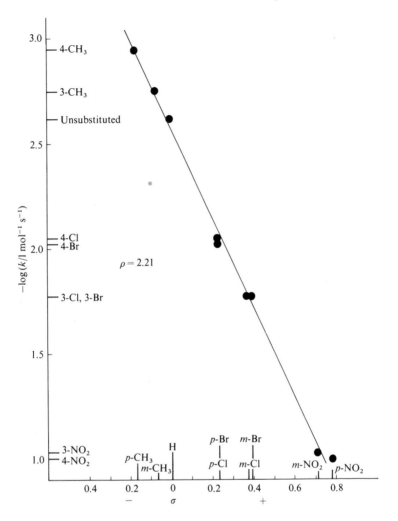

Fig. 4.8. The correlation of hydrolysis rates of 3- and 4-substituted ethyl 1-naph-thoates with σ_m and σ_p.

Structures (58) to (63) show some typical conjugated hydrocarbon structures, which for the purpose of development of the theory are divided into different types.

Butadiene (58), benzene (59), naphthalene (60), and the benzyl radical (61) are examples of alternant hydrocarbons (AH's), because the conjugated atoms can be starred so that no two starred or unstarred positions are joined. Azulene (62) and fulvene (63) are non-alternant because they cannot be starred in this way, and the PMO method cannot deal with such molecules. AH's are either even, with equal numbers of starred and unstarred positions (e.g. 58, 59 and 60) or odd (e.g. 61).

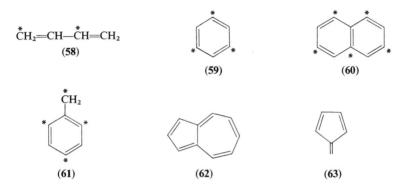

The molecular orbitals ψ_{MO} are obtained by Hückel MO theory using a linear combination of atomic p orbitals χ (AO's), the influence of the σ electrons on the energy of the AO's being neglected. This approximation is probably not serious; in any case it is only one of many involved in the simple Hückel method, which is thus hopelessly inaccurate for absolute assessment of molecular properties. It becomes, however, much more dependable when a series of molecules are compared for each one of which the approximations involved are similar and thus largely cancel, which is just the sort of situation to which we are seeking to apply it here.

For an even AH the MO's fall into pairs symmetrically disposed, one pair bonding (BMO) and the other anti-bonding (ABMO). In the case of an odd AH all but one of the MO's are also paired in this way, the odd MO having an energy equivalent to a carbon 2p atomic orbital. It thus contributes nothing to the overall binding energy and is termed therefore a non-bonding MO (NBMO). In an even AH the electrons occupy the bonding MO's in pairs, but for an odd AH radical, one electron occupies the NBMO. The situation is exemplified by fig. 4.9, in which the MO's are constructed for benzene and the benzyl cation, radical and anion.

The mixing coefficients for the AO's in a NBMO are readily found by the following rules.

1. Star the odd AH as described previously, *ensuring that the greater number of positions is starred.* Only the *p*-orbitals of these active

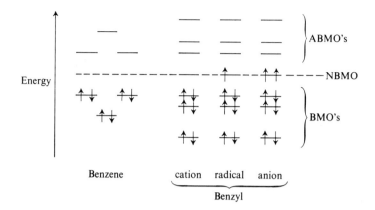

Fig. 4.9. MO's and their electron occupants in benzene and benzyl.

positions contribute to the NBMO; those that are unstarred are inactive and make no contribution.
2. The relative size of the coefficients is calculated from the fact that their sum for positions directly linked to an unstarred position is zero.
3. The absolute size of the coefficients is decided by the normalisation procedure – the sum of their squares is unity.

The simplicity of this process is revealed on application to specific examples. It is demonstrated in fig. 4.10.

Thus the NBMO for benzyl is given by

$$\psi_{NBMO} = -\frac{1}{\sqrt{7}}\chi_2 + \frac{1}{\sqrt{7}}\chi_4 - \frac{1}{\sqrt{7}}\chi_6 + \frac{2}{\sqrt{7}}\chi_7 \qquad (4.10)$$

The total π electron density in an even AH can be shown by Hückel MO calculations to be unity at all positions. This means that a side chain attached to any such position is predicted to undergo reaction at the same rate. Experiment bears this out as table 4.4 testifies. Gathered here are data for alkaline hydrolysis of methyl esters of arylcarboxylates (Corbett, Feinstein, Gore, Reed & Vignes, 1969), which show that, provided steric interactions are absent or constant, rates of hydrolysis are remarkably similar. Indeed, agreement is so good that it may be offered as a strong indication, if not a proof, that Hammett's device of measuring the relative electron density at a position in benzene or any other aromatic system, by attachment of a reacting side chain, is a very sound one.

1. Starring

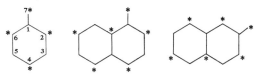

2. Calculation of coefficients

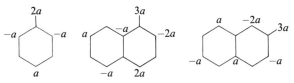

3. Calculation of a

$$a^2 + a^2 + a^2 + 4a^2 = 1 \qquad a^2 + a^2 + a^2 + 4a^2 + 4a^2 + 9a^2 = 1$$

$$a = 1/\sqrt{7} \qquad\qquad\qquad a = 1/\sqrt{20}$$

$$a^2 + a^2 + a^2 + a^2 + 4a^2 + 9a^2 = 1$$

$$a = 1/\sqrt{17}$$

Fig. 4.10. Calculation of NBMO coefficients for benzyl and α- and β-naphthyl-methylene carbanions.

The π electron density of an odd AH is given by the squares of the NBMO coefficients. The resultant charge densities of the carbanions, symbol q, are given in (64), (65) and (66) for the three systems in fig. 4.10.

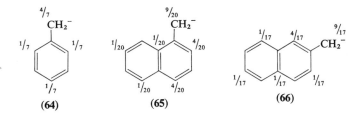

It should be noticed that since NBMO's neither add to nor subtract from the π electron interaction energy (fig. 4.9), calculations employing these orbitals may be applied to cationic, radical or anionic odd AH's with equal validity.

We are now in possession of sufficient background information to follow Dewar's novel method of calculating σ values for substituent effects on reaction rates in polycyclic derivatives (although this is only one use of this theory, which has been applied with considerable success

TABLE 4.4   *Rates of alkaline hydrolysis of methyl esters of aryl carboxylic acids* (70 % *aq. dioxan*, 40 °C)

Sterically unhindered esters		Sterically hindered esters	
Ester	$10^2 k/l$ $\text{mol}^{-1}\,\text{s}^{-1}$	Ester	$10^2 k/l$ $\text{mol}^{-1}\,\text{s}^{-1}$
Benzoate	2.79	1-Naphthoate (one peri-H)	1.07
2-Naphthoate	2.73	1-Anthroate (one peri-H)	0.82
2-Anthroate	3.67	1-Phenanthroate (one peri-H)	1.01
2-Phenanthroate	3.84	9-Phenanthroate (one peri-H)	1.46
3-Phenanthroate	3.09	9-Anthroate (two peri-H's)	0.06

to the general interpretation of reactivity and spectral data for polycyclic aromatic systems).

The equation for calculation of a given σ value is

$$\sigma = \frac{F}{r} + Mq \qquad (4.11)$$

The first term in (4.11) refers to the inductive effect. F is a measure of the field set up by the substituent and r is the distance between the points of attachment of side chain and substituent, the assumption being that the inductive effect is transmitted through space (D) rather than through the σ bonds (I_σ), although it seems certain that calculations using the distance round the σ bonds would generally give similar results. Values of r, taking the C—C bond length in benzene to be 1, and assuming all C—C bonds to have the same length, are given in table 4.5 for benzene, naphthalene and biphenyl.

The second term in equation (4.11) refers to the resonance effect, where q is the negative or positive charge, produced by substituent $\overline{CH}_2$ or $\overset{+}{CH}_2$ respectively, at the point of attachment of the reacting side chain. Values of q are included in table 4.5.

Values of F and M are found by interpolation in the benzene system. Thus for the methoxy substituent:

$$\sigma_m = 0.12 = F/\sqrt{3}, \qquad F = 0.20$$

$$\sigma_p = -0.27 = \frac{0.20}{2} + \frac{M}{7}, \qquad M = -2.58$$

TABLE 4.5   *Values of r and q for benzene, naphthalene and biphenyl*

Compound	Substituent position	Side chain position	r	q
Benzene	3	1 (*m*)	$\sqrt{3}$	0
	4	1 (*p*)	2	1/7
Naphthalene	3	1	$\sqrt{3}$	0
	4	1	2	1/5
	5	1	$\sqrt{7}$	1/20
	6	1	3	0
	7	1	$\sqrt{7}$	1/17
	4	2	$\sqrt{3}$	0
	5	2	3	0
	6	2	$\sqrt{13}$	1/17
	7	2	$\sqrt{12}$	0
	8	2	$\sqrt{7}$	1/20
Biphenyl	3'	1	$\sqrt{21}$	0
	4'	1	5	1/31

While for NO_2:

$$\sigma_m = 0.71 = F/\sqrt{3}, \qquad F = 1.23$$

$$\sigma_p = 0.78 = \frac{1.23}{2} + \frac{M}{7}, \qquad M = 1.14$$

Table 4.6 gives F and M for some common substituents, calculated in this way, and thence σ values for the naphthalene system from (4.11), while table 4.7 furnishes data from which the σ values may be tested, as

TABLE 4.6   *Values of F and M derived from σ_m and σ_p, and of σ for naphthalenes with a reacting side chain in the 1-position*

Substituent	F	M	σ_3	σ_4	σ_5	σ_6	σ_7
OCH_3	0.20	−2.58	0.12	−0.42	−0.05	0.07	−0.07
CH_3	−0.12	−0.77	−0.07	−0.21	−0.07	−0.04	−0.09
Cl	0.65	−0.70	0.37	0.19	0.21	0.23	0.21
Br	0.68	−0.77	0.39	0.19	0.22	0.23	0.21
NO_2	1.23	1.14	0.71	0.84	0.52	0.41	0.53

TABLE 4.7   *Reactivity data for substituent effects on 1-substituted naphthalenes* (Dewar & Grisdale, 1962; Price, Mertz & Wilson, 1954)

pK_a values of 1-naphthoic acids (50% aq. ethanol, 25 °C)					
Substituent	pK_a	Substituent	pK_a	Substituent	pK_a
H	5.54				
4-NO$_2$	4.23	4-Br	5.09	4-OCH$_3$	6.09
5-NO$_2$	4.72	5-Br	5.09	5-OCH$_3$	5.56
6-NO$_2$	4.92	6-Br	5.26	6-OCH$_3$	5.63
7-NO$_2$	4.99	7-Br	5.43	7-OCH$_3$	5.66

Rates of alkaline hydrolysis of ethyl 1-naphthoates (70% aq. dioxan, 25 °C)

Substituent	$10^3 k/\mathrm{l\ mol^{-1}\ s^{-1}}$
H	1.28
4-Cl	4.42
7-Cl	2.38
4-NO$_2$	46.5
5-NO$_2$	18.0

shown in figs. 4.11 and 4.12. Considering the approximations used in the calculations of the σ values, of which probably the most notable is that substituents attached to unstarred positions are assumed to have zero resonance interaction, whereas we know that in fact, from § 3.4, a small secondary effect is operative, agreement is quite reasonable. The ρ value for the pK_a value correlation is 1.5, while that for the ester hydrolyses is

TABLE 4.8   *Substituent effects in the detritiation of 1- and 2-tritio-naphthalenes*

1-Tritionaphthalene			2-Tritionaphthalene		
Substituent	$\log (k/\mathrm{s^{-1}})^a$	σ^+	Substituent	$\log (k/\mathrm{s^{-1}})^b$	σ^+
4-OCH$_3$	3.93	−0.95	4-OCH$_3$	0.20	0.05
5-OCH$_3$	0.59	−0.22	6-OCH$_3$	2.09	−0.27
3-Cl	−2.64	0.39	4-Cl	−1.99	0.39
4-Cl	−0.58	0.01			
5-Cl	−1.55	0.17			

[a] Relative to unsubstituted 1-tritionaphthalene
[b] Relative to unsubstituted 2-tritionaphthalene

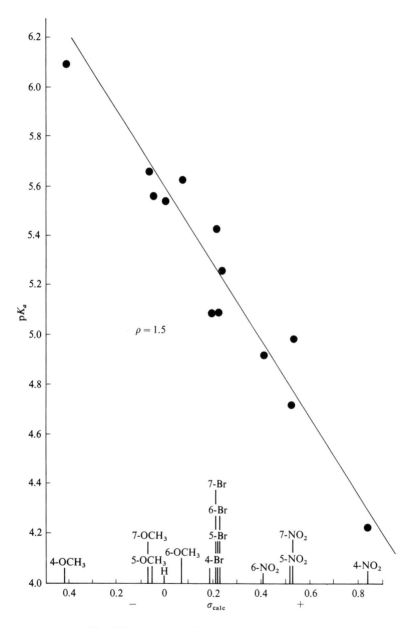

Fig. 4.11. pK_a values of 1-naphthoic acids *vs.* σ_{calc}.

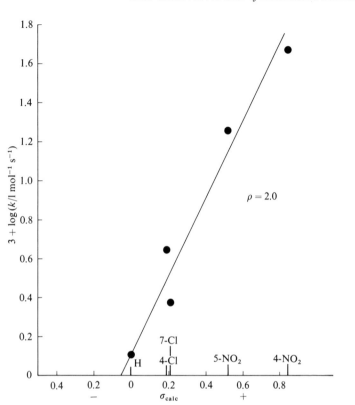

Fig. 4.12. Log k for rates of alkaline hydrolysis of ethyl 1-naphthoates *vs.* σ_{calc}.

2.0, which may be collated with 1.60 and 1.82, the respective values for the analogous benzene reactions.

In precisely analogous fashion, σ^+ and σ^- values may be partitioned into F and M components and thence applied to electrophilic or nucleophilic substitutions in polycyclic aromatic systems. Table 4.8 shows the calculations of σ^+ for OCH_3 and Cl for reactions at the 1- and 2-positions in naphthalene. In fig. 4.13, they are applied to detritiation of 1- and 2-tritionaphthalenes (Eaborn & Fischer, 1969) producing quite a reasonable correlation.

Another interesting application of NBMO theory to aromatic electrophilic substitution is worth enquiring into here, because the results are compatible with deductions reached using the Hammett equation.

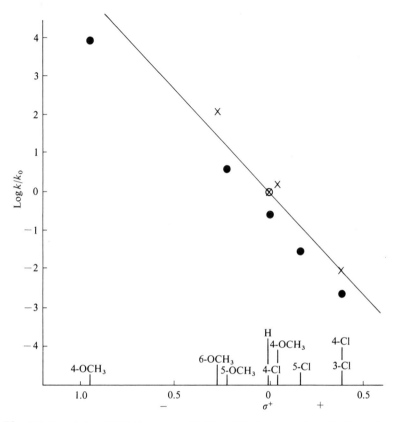

Fig. 4.13. Log k for detritiation rates of tritionaphthalenes *vs.* σ^+_{calc}. 1-tritionaphthalenes, ●; 2-tritionaphthalenes, ×.

Dewar (1969) has shown that the π electron energy change ΔE_π associated with the conversion of an even into an odd AH, such as shown in [*4.6*] for an electrophilic substitution process

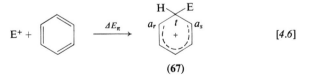

$$E^+ + \quad \xrightarrow{\Delta E_\pi} \quad \text{(67)} \qquad [4.6]$$

is given by equation (4.12)

$$\Delta E_\pi = -2\beta(a_r + a_s) = -\beta N_t \qquad (4.12)$$

Constant β, the C—C bond resonance integral, has a value of *ca.* -84 kJ mol^{-1}, while a_r and a_s are the NBMO coefficients of the sp^2 carbon atoms r and s of the odd AH (**67**) adjacent to the sp^3 carbon atom t, which has been attacked by electrophile E$^+$. N_t is known as a *reactivity number*, and is defined by (4.13)

$$N_t = 2(a_r + a_s) \qquad (4.13)$$

Values of N_t for various positions in the hydrocarbons benzene (**68**), naphthalene (**69**), biphenyl (**70**), anthracene (**71**) and phenanthrene (**72**) and indeed any AH however complex, are thus readily calculated according to the rules given previously. Figure 4.14 shows the stepwise process for 1-substitution in naphthalene and 2-substitution in phenanthrene. The student should make sure that he can compute other values for himself.

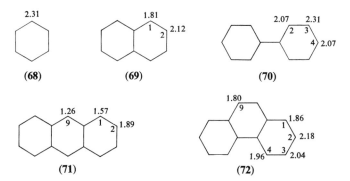

The bonding and solvation changes accompanying electrophilic substitution are essentially constant for all aromatics, with the exception of ΔE_π. ΔE_π and thus N_t may therefore reasonably be assumed proportional to the overall free energy change of the reaction; according to the nomenclature explained in chapter 5, it will constitute an internal enthalpy change.

Thus

$$2.3\,RT \log f_t = -\beta(N_t - N_B) \qquad (4.14)$$

where f_t is the partial rate factor for electrophilic substitution in one position, t, of an AH, and N_B is the reactivity number for benzene, which is 2.31. Plots of $\log f_t$ *vs.* N_t are thus predicted to be straight lines of slope $-\beta/2.3\,RT$. Approximately straight lines are indeed obtained, but their slopes yield values of β which are considerably less negative than

(a)

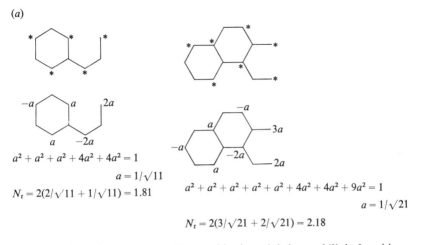

$$a^2 + a^2 + a^2 + 4a^2 + 4a^2 = 1$$

$$a = 1/\sqrt{11}$$

$$N_t = 2(2/\sqrt{11} + 1/\sqrt{11}) = 1.81$$

$$a^2 + a^2 + a^2 + a^2 + a^2 + 4a^2 + 4a^2 + 9a^2 = 1$$

$$a = 1/\sqrt{21}$$

$$N_t = 2(3/\sqrt{21} + 2/\sqrt{21}) = 2.18$$

Fig. 4.14. Calculation of N_t for (a) the 1-position in naphthalene and (b) the 2-position in phenanthrene.

-84 kJ mol^{-1}. Figure 4.15 shows the correlation obtained for bromination in acetic acid at 25°, which yields a value of β of -67 kJ mol^{-1} (Altschuler & Berliner, 1966).

This deviation is explained by consideration of [*4.6*], where the model for the transition state is taken as the Wheland intermediate; in fact, however, the transition state will be only partway along the reaction co-ordinate, between the reactant molecule and the intermediate, and the hybridisation of the carbon atom t is a mixture of the sp^2 and sp^3 forms. The observed β value is thus a measure of the selectivity of the electrophile concerned, being the less negative the more the transition state resembles the ground state. It should be therefore proportional to the ρ value for the same electrophile as evaluated from its reaction with monosubstituted benzenes (§ 2.7). Comparison of observed β values (Altschuler & Berliner, 1966; Streitwieser, 1962; Norman & Taylor, 1965; Streitwieser, Mowery, Jesaitis & Lewis, 1970; Olah, 1971) with ρ values, as given in table 4.9, confirms to a reasonable degree of reliability the theoretical reasoning. Basicity of AH's in hydrofluoric acid, an equilibrium rather than a kinetic process, involving conjugate acids of σ complex (Wheland intermediate) type, gives as expected the closest correspondence of β_{obs} to -84 kJ mol^{-1}.

As a final illustration of the correspondence of the conclusions of NBMO theory, to those derived from the Hammett equation, we

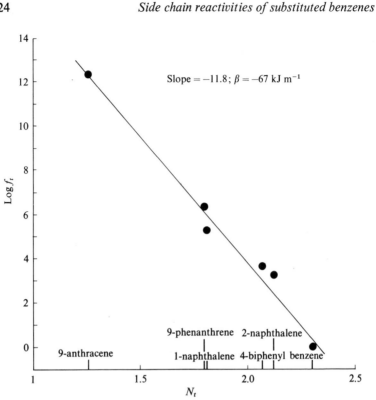

Fig. 4.15. Log partial rate factors *vs.* N_t for bromination.

consider the solvolysis of 2-arylethyl *p*-toluene sulphonates (Bentley & Dewar, 1970). In § 2.8, it was found that electron donor groups in the aryl (benzene) nucleus induced the formation of a phenonium cation (**40**) in an S_N1 process, the resultant log rate constants being correlated by σ^+.

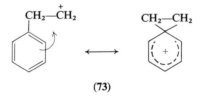

(**73**)

By analogous reasoning, if the aryl group is an AH, the degree to which it encourages the S_N1 process should be proportional to the reactivity number N_t corresponding to the point of attachment of the reacting side chain, since the formation of cation (**40**) or (**73**) from the ground state

TABLE 4.9 *Correspondence between ρ and β_{obs} values*

Electrophilic substitution reaction	$-\rho$	$-\beta_{obs}/\text{kJ mol}^{-1}$
Basicity (HF, 25 °C)	14.0	79
Bromination (CH_3COOH, H_2O, 25 °C)	12.1	67
Chlorination (CH_3COOH, 25 °C)	10.0	50
Protodetritiation (CF_3COOH, 70 °C)	8.2	42
Deuterodeprotonation (CF_3COOD, D_2SO_4, CCl_4, 30 °C)		38
Nitration (($CH_3CO)_2O$, 25 °C)	6.0	25
Protodesilylation ($HClO_4$, CH_3OH, H_2O, 50 °C)	4.6	13

aromatic sulphonate is essentially equivalent, in terms of the π electron energy change, to that of formation of the Wheland intermediate from the AH. Table 4.10 affords data for the sulphonate solvolysis in formic, acetic and trifluoracetic acids for which the correlation of the log rate constants with N_t can be examined, as shown in fig. 4.16.

Generally, nucleophilicity is not equatable with acidity, but in this instance, where the nucleophile is in all cases the carboxyl group, the two concepts correspond in that the greater the tendency of the carboxyl group to ionise, the less is its tendency to attack a carbon atom in an S_N2 style mechanism.

The figure demonstrates the transition from the S_N2 to the S_N1 'anchimerically assisted' reaction in the case of acetic (pK_a 4.8) and formic acid (pK_a 3.8), in convincing support of Schleyer's conclusions (§ 2.8). In addition, it demonstrates how the least reactive nucleophile trifluoroacetic acid (pK_a −0.3) sponsors only the S_N1 mode.

TABLE 4.10 *Log relative rate constants for the solvolysis of 2-arylethyl p-toluene sulphonates*, Ar CH_2CH_2OTs

		log k_{rel}		
Aryl group	N_t	CF_3COOH (55 °C)	HCOOH (75 °C)	CH_3COOH (75 °C)
Phenyl	2.31	0	0	0
9-Anthryl	1.26		1.94	1.61
1-Naphthyl	1.81	0.88	0.60	0.13
2-Naphthyl	2.12	0.30	0.12	0.04
4-Biphenylyl	2.07			0.06
9-Phenanthryl	1.80		0.50	
1-Pyrenyl	1.51	1.48	1.06	0.84

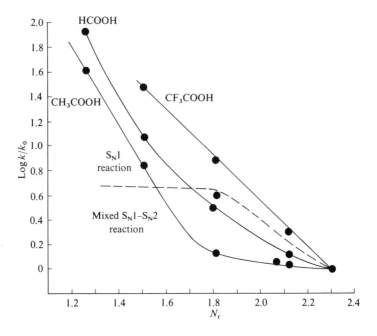

Fig. 4.16. Solvolysis of 2-arylethyl *p*-toluene sulphonates.

4.9. Application of the Hammett equation to spectral measurements. We have given considerable thought to the application of Hammett style equations to the energy differences between two molecular ground states (equilibria), or between ground and transition states (reactions). Certainly these considerations represent the most extensive use of such equations, but it is of interest to enquire to what extent they will correlate spectral characteristics of ground state molecules. Such molecules are not of course static, but exist in a state of dynamic equilibrium undergoing rotational and vibrational transitions which require somewhat lower energies than are generally necessary for bond breaking and making processes in reaction, although this depends very much on the type of transition involved. Thus the energies associated with vibrational frequencies, observed in the infra-red (IR) region of the electromagnetic spectrum, are of the order of 10 kJ mol^{-1}, but are many million times smaller than this for the nuclear transitions implicated in nuclear magnetic resonance (NMR). In spectral techniques we encourage, by supplying suitable quanta, all the molecules to undergo energy transitions which in the normal situation, at any given instant, only a proportion would be experiencing.

Rao (1963) has provided a number of examples of the correlation of the frequencies of vibrations in side chain functions, such as the carbonyl stretch in acetophenones and benzoic acids and esters, with σ values of substituents in the aromatic nucleus. The intensities of such vibrations also may afford correlations, and thus, since the intensity of an IR band is proportional to the dipole moment change during the vibration, one might therefore expect a corresponding correlation with dipole moment measurements. However, quantitative relationships in this case are limited, because such measurements represent a composite quantity incorporating the direction and magnitude of all electronic polarisations within the molecule. No doubt if individual bond moments could be separated out, better correspondence would be achieved.

A series of significant IR measurements have been made by Katritzky and Topsom and their co-workers (Brownlee, Hutchinson, Katritzky, Tidwell & Topsom, 1968). Their procedure involves measurements of the integrated area A for the vibration of frequency about 1600 cm^{-1} in monosubstituted benzenes. This vibration is believed to involve a variation in resonance interaction between substituent and ring. This is borne out by the fact that band intensities A are quantitatively connected with σ^0_R for the substituent by equation (4.15)

$$A = 17600 \, (\sigma^0_R)^2 + 100 \qquad (4.15)$$

which has subsequently been used to amass a comprehensive list of σ^0_R values. Equation (4.15) does not allow determination of the sign of σ^0_R, but this of course is generally apparent by qualitative reasoning. Table

TABLE 4.11 σ^0_R *values*

Substituent	$\sigma^0_R{}^a$	$(\pm)\sigma^0_R{}^b$
NH$_2$	-0.48	0.47
OCH$_3$	-0.35	0.43
Cl	-0.20	0.22
CH$_3$	-0.10	0.10
COOC$_2$H$_5$	0.14	0.18
NO$_2$	0.18	0.17
$\overset{+}{N}(CH_3)_3$		0.15

[a] From table 3.6
[b] From equation (4.15)

TABLE 4.12 σ_I values

Substituent	$\sigma_I{}^a$		$\sigma_I{}^b$		
			Neutral solvents[c]	Weakly protonic solvents	Trifluoro-acetic acid
$N(CH_3)_2$	0.10,	0.04	0.10	0.10	
OCH_3	0.28,	0.25	0.25	0.29	0.51
CH_3	−0.01,	−0.05	−0.08	−0.08	−0.06
Br	0.45,	0.45	0.44	0.44	0.44
$COCH_3$	0.27,	0.28	0.18	0.23	0.46
$COOC_2H_5$	0.29,	0.30	0.11	0.21	0.35
CN	0.56,	0.60	0.48	0.53	0.74
NO_2	0.64,	0.63	0.56	0.60	0.80
$\overset{+}{N}(CH_3)_3$	0.91,	0.90		0.93	

[a] Various values from tables 3.1 and 3.3, or calculated by equations (3.10) and (3.11)
[b] From Taft's ^{19}F shielding experiments
[c] Cyclohexane, benzene, acetone, carbon tetrachloride etc.

4.11 gives some results obtained by this procedure, using carbon tetrachloride as solvent, together with $\sigma^0{}_R$ values obtained by application of the Taft equation (3.11) to σ^0 values, and previously given in table 3.6.

The values obtained for CH_3, C_2H_5, $CH(CH_3)_2$, $C(CH_3)_3$ and $\overset{+}{N}(CH_3)_3$, (−) 0.10, 0.10, 0.12, 0.15 and 0.15, provide further evidence for the existence of C—H, C—C and N—C hyperconjugation. Another interesting feature of these measurements is that the $\sigma^0{}_R$ values thus obtained show no variation when the solvent is changed to cyclohexane, chloroform or isopropanol. This affords convincing confirmation of the conclusions reached in § 1.6, where it was decided that in general σ values are not susceptible to solvation influences, particularly when one remembers that $\sigma^0{}_R$ measures resonance effects which are of high polarisability and thus sensitive to differences in solute–solvent interactions if such differences existed.

Taft (Taft, Price, Fox, Lewis, Andersen & Davis, 1963) has investigated the use of the magnetic shielding of the ^{19}F nucleus in the benzene ring as a measure of the inductive effect of a substituent m to it. Such shieldings, $\int_H^{m-x}$, are found to reflect, by equation (4.16), solely the inductive effect of the m-substituent

$$\int_H^{m-x} = 7.10\,\sigma_I + 0.60 \qquad (4.16)$$

Having established the authenticity of equation (4.16) by use of known σ_I values obtained as described in chapter 3, Taft went on to utilise the equation for measurement of σ_I for a large number of groups in a variety of solvents. Some representative results are given in table 4.12.

Disconcertingly, there is seen to be some solvent-induced variation, a surprising result in view of previous conclusions, particularly as the inductive effect is normally thought of as having a low polarisability. However, the variation, although outside experimental error is only large for the solvent trifluoroacetic acid. This is a very strong acid, and pronounced interactions of the form of (74) and (75) may be postulated for groups such as NO_2 and OCH_3, or for carbonyl-containing groups

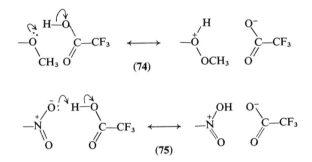

In confirmation, σ_I values of substituents Br and CH_3, for which clearly there can be no interactions of this form, show no variation in this solvent from values in other solvents.

Another set of revealing NMR measurements have been made by Williamson (Williamson, Jacobus & Soucy, 1964), who examined the chemical shifts of H_{exo} and H_{endo} in structure (76), produced by change of substituent X

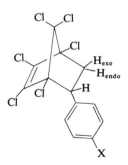

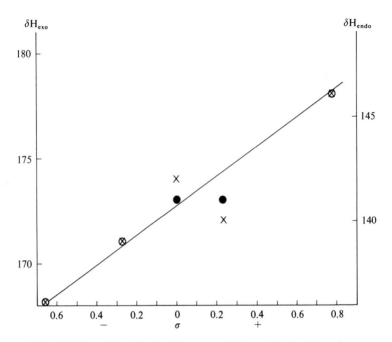

Fig. 4.17. Exo- (×) and endo- (●) proton shifts correlated with σ values.

The effect of X on the electronic shielding of H_{exo} and H_{endo} is conveyed by induction and resonance to the point of attachment of the benzene nucleus to the norbornenyl system, and thence via the σ framework, when H_{exo} and H_{endo} should experience the same I effect, or the inductive polarisation of the C—X bond may be communicated by the direct field effect D, which would influence the electronic shielding of H_{endo} more than H_{exo}. As fig. 4.17 shows, the chemical shifts of H_{exo} and H_{endo}, measured in Hz downfield from the tetramethylsilane signal at 60 MHz in carbon disulphide solvent, are within reasonable accuracy equal to one another, and proportional to σ_p. This speaks strongly for the I_σ mode of transmission of the inductive effect in this case, and would in general imply that both forms, rather than D alone, must be taken into account.

4.10. Problems

20. The condensation products of *o*-hydroxybenzylamine with substituted benzaldehydes are Schiff bases in equilibrium with ring tautomers.

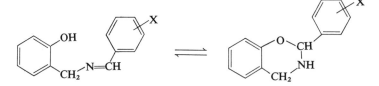

The following table gives the % ring compound in the tautomeric equilibria established in chloroform-*d*, and the chemical shifts of the hydroxyl proton in the open chain form with tetramethylsilane as internal standard. Examine the applicability of the Hammett equation to these measurements.

X	% ring form	Chemical shift of OH proton (δ, Hz)
p-NO$_2$	51	8.92
m-NO$_2$	49	9.00
p-Br	20	9.67
H	16	9.97
p-CH(CH$_3$)$_2$	10	10.15
p-N(CH$_3$)$_2$	*undetectable*	10.17

(McDonagh & Smith, 1968)

21. UV measurements show equal amounts of 3-hydroxypyridine and its zwitterionic tautomer in water at 25 °C. Taking ρ for the ionisation of phenols and pyridinium ions as 2.1 and 6.0 respectively, the pK_a's of pyridine and phenol as 5.21 and 9.99 respectively, calculate the effective σ_m value for azonium in this medium (Bryson, 1960).

22. Calculate σ values for the substituents in the polycyclic systems for which reactivity data are given below, and thus comment on the correlations obtained.

pK_a *values of biphenyl 4-carboxylic acids* (50 % *aq. butyl cellosolve*, 25 °C)

4'-NO$_2$	5.25	H	5.66
3'-NO$_2$	5.35	4'-OCH$_3$	5.75
4'-Br	5.47		

(Berliner & Blommers, 1960)

pK_a *values of 2-naphthoic acids (50%*
aq. ethanol, 25 °C) and log rate constants
for alkaline hydrolysis of methyl 2-naph-
thoates (70% aq. dioxan, 25 °C)

	pK_a	$3 + \log (k/\text{l mol}^{-1} \text{s}^{-1})$
6-NO$_2$	4.98	2.57
7-NO$_2$	5.15	2.16
6-Cl	5.42	1.67
7-Cl	5.39	1.48
H	5.66	1.09
6-OCH$_3$	5.82	0.76
7-OCH$_3$	5.67	

(Wells, Ehrenson & Taft, 1968)

23. From data given in § 4.3, calculate pK_1 for scheme [*4.2*] and hence evaluate σ_p for COO$^-$.

24. The rate constants $(10^4 k/\text{l mol}^{-1} \text{s}^{-1})$ for alkaline hydrolysis of 5-substituted 2-methoxytropones (40% aq. dioxan, 30 °C) are as follows:

H	2.50	Br	52.7
OCH$_3$	0.670	N=NC$_6$H$_5$	945
CH(CH$_3$)$_2$	0.655	NO$_2$	842.00
Cl	34.5		

(Bowden & Price, 1971)

Investigate the application of the Hammett equation to these results.

25. The pK_a values of 4-substituted quinoline 2-carboxylic acids (44% aq. ethanol, °25 C) are as follows:

4-aza	3.69	H	4.95
4-Cl	4.28	6-CH$_3$	5.14
4-OCH$_3$	6.29	8-NO$_2$	4.15

(Donaldson & Joullié, 1968)

All the acids, with the exception of that with the 4-OCH$_3$ substituent, showed an absorption band in the 1725–1680 cm^{-1} region. What is the predominant tautomeric form of the acids?

5 Thermodynamic aspects of the Hammett equation

5.1. Introduction. So far the Hammett equation has been considered as an empirical relationship, with the arbitrary acceptance that free energy changes, and thus σ and ρ values, may be rationalised in terms of the electronic interactions of substituents and reaction centres.

For a single molecular species at absolute zero in a vacuum, the free energy of the molecule, which is equivalent to its enthalpy, derives from zero point vibrations, together with potential energy contributions described by inductive and resonance effects. These effects are the only factors utilised in explaining the magnitude and sign of substituent and reaction constants. At finite temperatures, under the influence of solvent molecules, the potential energy terms will be further overlaid by other kinetic energies distributed among the various degrees of freedom of the molecule, together with kinetic and potential energy contributions associated with solvation. The total free energy is consequently an intricate blend of many energy and probability factors, composing the total enthalpy and entropy, and the changes in these quantities when the system undergoes reaction are not easy to predict with theoretical precision. It is therefore appropriate to consider, at least briefly, the thermodynamic implications of the use of the Hammett equation. This is a subject of doubt and dispute, but we can at least attempt to discern the relevance of the main ideas suggested, and the degree to which they accord with the experimental data.

Moreover, it is worth noticing that the correlation of the equation with thermodynamic parameters is part of a much larger question. Logarithmic rates, and therefore free energies of activation, of reactions in general are satisfactorily explained by simple electronic theory. However, the apparently more relevant enthalpies, which contain the total contribution from potential energies of inductive and resonance form, often show variations which can in no way be rationalised in the same simple terms. This is true for reaction series where the only difference between one compound and another in the set is a change in substitution well removed from the reaction site; *although it is sometimes*

133

*stated and frequently implied, that reactions which follow the Hammett
relationship have ΔH values accurately proportional to ΔG, and ΔS values
zero or constant, this is by no means invariably or necessarily the case.*

Attention will thus be paid in this final chapter to the discussion of
this problem, although as yet no simple explanation can be given. Moreover, in referring to it as a problem, we take the viewpoint of the physical
chemist. It is really a source of satisfaction to the organic chemist, since
it not merely enables but actually encourages him, in mechanistic
correlations, to restrict his attention to single temperature measurements, readily obtained by comparison with the high accuracy multiple
temperature experiments necessary for elucidation of reliable ΔH and
ΔS terms.

The point may be illustrated by three examples. The first is the
ionisation of carboxylic acids. In table 5.1 are recorded thermodynamic
parameters for the ionisation of substituted formic acids (Larson &
Hepler, 1969). The pK_a values and resultant ΔG^0 values are obviously
explicable in terms of relative stabilisation of the carboxylate anions by
the inductive effects of the various substituents, steric effects being
negligible. In fact, a linear plot of ΔG^0 against σ^* is obtained (fig. 5.1).

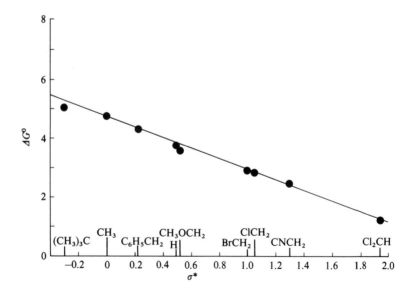

Fig. 5.1. Correlation of ΔG^0 for ionisation of substituted formic acids with σ^*.

TABLE 5.1 *Thermodynamic parameters for ionisation of some substituted formic acids*, XCOOH (H$_2$O, 25 °C)

X	pK_a	ΔG^0/4.18 kJ mol^{-1}	ΔH^0/4.18 kJ mol^{-1}	ΔS^0/4.18 J K^{-1} mol^{-1}	σ^*
Formic acids					(Tables 3.3 and 3.4)
H	3.77	5.12	0	−17.2	0.49
CH$_3$	4.76	6.48	−0.10	−22.1	0.00
BrCH$_2$	2.92	3.96	−1.24	−17.4	1.00
ClCH$_2$	2.88	3.91	−1.14	−16.9	1.05
NCCH$_2$	2.48	3.37	−0.89	−14.3	1.30
Cl$_2$CH	1.25	1.7	−0.1	−6	1.94
C$_6$H$_5$CH$_2$	4.33	5.88	−0.88	−22.7	0.22
(CH$_3$)$_3$C	5.05	6.86	−0.72	−25.4	−0.30
CH$_3$OCH$_2$	3.52	4.87	−0.96	−19.6	0.52

However, the variation of ΔH^0 and ΔS^0 in no way indicates a simple dependence on substituent effects, as plots of ΔG^0 vs. ΔH^0 and ΔS^0 demonstrate (figs. 5.2 and 5.3). ΔG^0 values for benzoic acids have also shown no correlation with ΔH^0 or ΔS^0 (Larson & Hepler, 1969), although a very recent preliminary report dealing with a limited number of examples suggests a correspondence does exist (Bolton, Fleming & Hall, 1972).

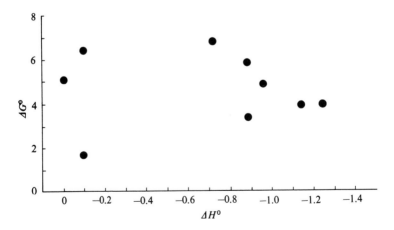

Fig. 5.2. ΔG^0 vs. ΔH^0 for ionisation of formic acids.

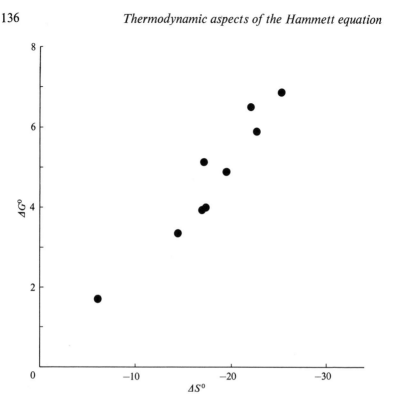

Fig. 5.3. ΔG^0 *vs.* ΔS^0 for ionisation of formic acids.

Rogne (1970) has examined the catalysis of benzenesulphonyl chloride hydrolysis by substituted pyridines.

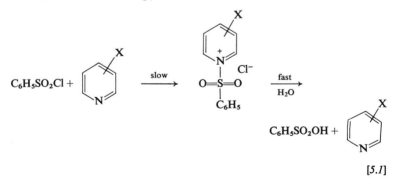

[5.1]

An excellent correlation of the logarithms of the second order rate constants against σ is obtained (fig. 5.4), provided σ_I is used for 4-COCH$_3$

and 4-CN, compatible with the results of similar pyridine–pyridinium ion reactions (§ 4.2). But the linearity of a plot of log k, and thus $\Delta G^{\ddagger}$, from equation (5.9), against $\Delta H^{\ddagger}$ is seen from fig. 5.5 to be of very low precision.

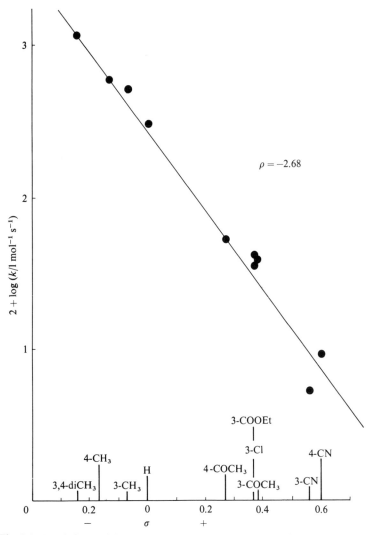

Fig. 5.4. Log k for pyridine-catalysed benzenesulphonyl chloride hydrolysis *vs.* σ (σ_I for 4-CN and 4-CH$_3$CO).

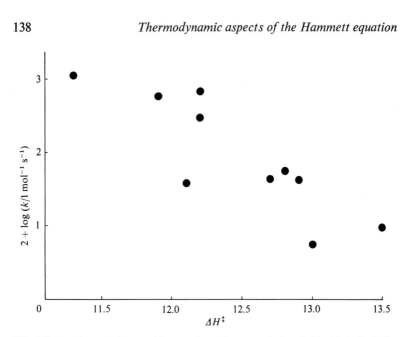

Fig. 5.5. Log k *vs.* $\Delta H^{\ddagger}$ for pyridine-catalysed benzenesulphonyl chloride hydrolysis.

In the third example, a plot of log k against $\Delta H^{\ddagger}$ for the S_N1 hydrolysis of phenyldimethylcarbinyl chlorides, the defining system for σ^+ constants, also shows poor correlation (fig. 5.6).

Before proceeding to a general analysis of effects for which these are specific examples, it is worthwhile reminding ourselves of the fundamental thermodynamic equations for reactions and equilibria.

5.2. Basic thermodynamic formulae. The Gibbs free energy change for an equilibrium defined by constant K is given by

$$\Delta G^0 = -RT \ln K = -2.3\ RT \log K \tag{5.1}$$

The superscript zero indicates that the function applies to species in an appropriate standard state, usually infinite dilution in pure water at 25 °C. If the variation of K with temperature has been experimentally determined, the enthalpy change for the equilibrium, ΔH^0, may be calculated from the van't Hoff equation

$$\Delta H^0 = -T^2 \frac{\mathrm{d}}{\mathrm{d}T}\left(\frac{\Delta G^0}{T}\right) = 2.3\ RT^2 \frac{\mathrm{d}\log K}{\mathrm{d}T} \tag{5.2}$$

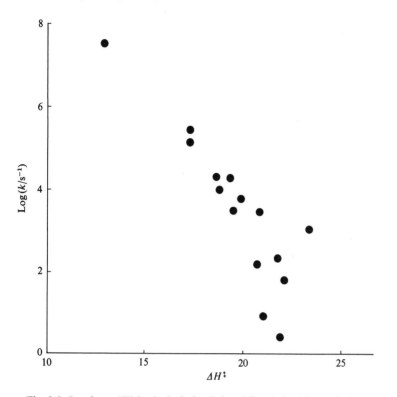

Fig. 5.6. Log k *vs.* $\Delta H^{\ddagger}$ for hydrolysis of phenyldimethylcarbinyl chlorides.

which assumes that there is no change in thermal capacity, that is $\Delta C_p = 0$.

Since

$$\Delta G^0 = \Delta H^0 - T \Delta S^0, \tag{5.3}$$

then

$$\Delta S^0 = -\frac{\mathrm{d}\,\Delta G^0}{\mathrm{d}T} = \frac{\Delta H^0 - \Delta G^0}{T} \tag{5.4}$$

An equivalent set of parameters may be derived for a reaction by invoking transition state theory. In this approach, the reactant molecules A and B are regarded on reaction in the rate determining step as combining to form an activated complex AB‡, whose energy is a maximum in the free energy profile:

$$A + B \quad \underset{}{\overset{K^\ddagger}{\rightleftharpoons}} \quad \underset{\substack{\text{activated} \\ \text{complex} \\ \text{or transition state}}}{AB^\ddagger} \quad \overset{k^\ddagger}{\longrightarrow} \quad \text{primary products}$$

$$\overset{\text{fast}}{\longrightarrow} \quad \text{secondary products}$$

The molecules A, B and AB‡ are considered to be in equilibrium, defined by the constant K where

$$K^\ddagger = \frac{[AB^\ddagger]}{[A][B]} \tag{5.5}$$

$k^\ddagger$ is the specific rate constant for decomposition of AB‡ to yield the products of reaction.

We may therefore write

$$\text{rate of reaction} = k_{obs}[A][B] = k^\ddagger[AB^\ddagger]$$

i.e.,

$$k_{obs} = k^\ddagger K^\ddagger \tag{5.6}$$

where k_{obs} is the observed rate constant for the reaction.

The activated complex is pictured as a normal molecule, stable in all degrees of freedom except one, which is the vibrational mode resulting in disruption of the complex leading to the primary reaction products. The frequency of this vibration is given by $\mathbf{k}T/\mathbf{h}$, where $\mathbf{k}$ and $\mathbf{h}$ are Boltzmann's (1.3805×10^{-23} J K^{-1}) and Planck's constant (6.626×10^{-34} J s) respectively. This represents the rate constant $k^\ddagger$ for decomposition of complex, which is thus constant for all reactions. Therefore

$$k_{obs} = \frac{\mathbf{k}T}{\mathbf{h}} K^\ddagger \tag{5.7}$$

A transmission coefficient κ may be inserted at this stage

$$k_{obs} = \kappa \frac{\mathbf{k}T}{\mathbf{h}} K^\ddagger \tag{5.8}$$

to account for the possibility that some activated complexes may not decompose to give products, but may revert to A and B. However, κ is generally considered to be unity for polar reactions in solution.

We can now set out the following parameters for reaction kinetics, analogous to those for equilibria already given:

$$\Delta G^{\ddagger} = -RT \ln K^{\ddagger} = -2.3 \, RT \log k_{obs} \, (h/kT) \qquad (5.9)$$

from which $\Delta G^{\ddagger}$ may be calculated. $\Delta G^{\ddagger}$ is thus proportional to $\log k$. From equation (5.7),

$$\ln k_{obs} = \ln kT/h + \ln K^{\ddagger}$$

and, since

$$\Delta G^{\ddagger} = \Delta H^{\ddagger} - T \Delta S^{\ddagger} \qquad (5.10)$$

$$\ln k_{obs} = \ln k/h + \ln T - \Delta H^{\ddagger}/RT + \Delta S^{\ddagger}/R \qquad (5.11)$$

In practice, values of the rate constant, k_{obs}, are measured at several different temperatures, and the Arrhenius activation energy E calculated from the equations:

$$\ln k_{obs} = \ln A - (E/RT),$$

where A is termed the frequency factor, so that

$$\frac{d \ln k_{obs}}{dT} = \frac{E}{RT^2} \qquad (5.12)$$

Differentiating (5.11) with respect to temperature, and substituting in equation (5.12), one obtains:

$$\frac{E}{RT^2} = \frac{\Delta H^{\ddagger} + RT}{RT^2}$$

$$E = \Delta H^{\ddagger} + RT \qquad (5.13)$$

and so $\Delta H^{\ddagger}$ and $\Delta S^{\ddagger}$ can be obtained from equations (5.10) and (5.13).

In this way, the thermodynamics of both equilibria and reactions may be treated in the same fashion. We should note, however, that thermodynamic parameters for equilibria are based on measurements on systems in the ground state under standard conditions, independent of the pathway by which the equilibria are set up, and are consequently amenable to very precise definition. On the other hand, such parameters for reactions involve the assumption of an activated complex, in equilibrium with reactant molecules, to which can be ascribed the properties of a ground state molecule proceeding to primary products with a transmission coefficient of unity, and thus their degree of validity is more doubtful.

5.3. The Hammett equation: a linear free energy relationship. We may
write, for a reaction correlated by the Hammett equation, that

$$\log \frac{k_X}{k_Y} = \rho \log \frac{K_X}{K_Y} \tag{5.14}$$

where k_X and k_Y are the rate constants for the side chain reaction of a
benzene nucleus bearing substituents X and Y in, say, the *p*-position,
while K_X and K_Y represent the ionisation constants of the equivalently
substituted benzoic acids in water at 25 °C. Then, from equations (5.1),
(5.9) and (5.14)

$$-\frac{\Delta G^{\ddagger}{}_X}{RT} + \frac{\Delta G^{\ddagger}{}_Y}{RT} = \rho \left(-\frac{\Delta G^{0}{}_X}{RT} + \frac{\Delta G^{0}{}_Y}{RT} \right)$$

so that

$$\Delta G^{\ddagger}{}_Y - \Delta G^{\ddagger}{}_X = \rho (\Delta G^{0}{}_Y - \Delta G^{0}{}_X)$$

which may conveniently be abbreviated to

$$\delta_X \Delta G^{\ddagger} = \rho \, \delta_X \Delta G^{0} \tag{5.15}$$

where δ_X represents the effect of a change in substitution. The Hammett
equation is therefore an example of a linear free energy relationship, a
term often abbreviated to LFER.

Now K_X/K_Y is the equilibrium constant of the so-called symmetrical
reaction:

$$X-C_6H_4-COOH + Y-C_6H_4-COO^- \xrightarrow{\ K_X/K_Y\ }$$

$$X-C_6H_4-COO^- + Y-C_6H_4-COOH$$

[5.2]

for which,

$$-2.3 \, RT \log K_X/K_Y = \delta_X \, \Delta G^{0}$$

Writing this out in terms of the free energies of the participant molecules,
we obtain

$$-2.3 RT \log K_X/K_Y = \delta_X \, \Delta G^{0} = G^{0}{}_{X-C_6H_4-COO^-} + G^{0}{}_{Y-C_6H_4-COOH}$$
$$- G^{0}{}_{X-C_6H_4-COOH} - G^{0}{}_{Y-C_6H_4-COO^-} \tag{5.16}$$

The component free energies of a term such as $G^0_{X-C_6H_4-COO^-}$ may be expressed by

$$G^0_{X-C_6H_4-COO^-} = G^0_X + G^0_{C_6H_4} + G^0_{COO^-} + G^0_{(X)(C_6H_4)}$$
$$+ G^0_{(C_6H_4)(COO^-)} + G^0_{(X)(COO^-)} \qquad (5.17)$$

where the first three terms are the free energies of the side chain, the communicating benzenoid system, and the substituent, in 'isolation', while the final three terms represent the free energies of interaction between these molecular components. If all four terms of the right-hand side of equation (5.16) are divided in this way, a sixteen-term expression is obtained, of which twelve terms cancel, yielding equation (5.18),

$$-2.3\,RT \log K_X/K_Y = \delta_X \, \Delta G^0 = G^0_{(X)(COO^-)} + G^0_{(Y)(COOH)} - G^0_{(X)(COOH)}$$
$$- G^0_{(Y)(COO^-)} \qquad (5.18)$$

so that the equilibrium constant ratio is seen to depend solely on the free energies of interaction between the carboxyl group and its anion with substituents X and Y in the p-position.

If the case of the reaction series is now considered, in which a side chain Z yields a transition state $Z^{\ddagger}$, then the symmetrical reaction for definition of $\log k_X/k_Y$ (5.13) is

$$X-C_6H_4-Z + Y-C_6H_4-Z^{\ddagger} \underset{}{\overset{k_X/k_Y}{\rightleftharpoons}} X-C_6H_4-Z^{\ddagger} + Y-C_6H_4-Z \qquad [5.3]$$

whence, by dissection analogous to (5.18), (5.19) is obtained.

$$-2.3\,RT \log k_X/k_Y = \delta_X \, \Delta G^{\ddagger} = G_{(X)(Z\ddagger)} + G^0_{(Y)(Z)} - G^0_{(X)(Z)}$$
$$- G^0_{(Y)(Z\ddagger)} \qquad (5.19)$$

Substitution of (5.18) and (5.19) into (5.15) thus demonstrates that the Hammett law depends for its success on the fact that the Gibbs free energy change ($\delta_X \, \Delta G^0$) produced by introduction of m- or p-substituents into the benzoic acid–anion system will be proportional to the corresponding change ($\delta_X \, \Delta G^0$ or $\delta_X \, \Delta G^+$) produced by the same substitution in a benzene ring bearing any other reacting side chain. If such a proportionality does not hold for certain substituents, we alter the character of the symmetrical reaction (5.2) to regain correspondence; hence the use of the anilinium–aniline or phenyldimethylcarbinyl chloride–carbonium ion system to match the character of substituent $-Z$ and $-Z^{\ddagger}$ interaction difference, as warranted by the circumstances discussed in chapter 2.

Most important, *viewing the Hammett equation in terms of symmetrical equilibria of the form of* [5.2] *and* [5.3] *makes it clear, from equations* (5.18) *and* (5.19), *that many of the complicated factors contributing to total free energy detailed in* § 5.1 *will cancel out, in that they are equivalent on both sides of the equilibria.*

5.4. The isokinetic relationship. Attention can now be turned to our initial enquiry – to what extent can LFER's be further rationalised into linear enthalpy relationships.
Since

$$\Delta G = \Delta H - T\Delta S,$$

$$\delta_X \Delta G = \delta_X \Delta H - T\delta_X \Delta S, \qquad (5.20)$$

when

$$\delta_X \Delta S = 0, \qquad \delta_X \Delta G = \delta_X \Delta H$$

Such a series is said to be *isoentropic*, and certain reaction series do correspond approximately to this pattern. The ionisation of anilinium ions, the defining system for σ^-, follows it exactly (Bolton & Hall, 1969). Another example, the alkaline hydrolysis of ethyl benzoates in 56 % aqueous acetone (Tommila & Hinshelwood, 1938), is given in table 5.2; obviously $\Delta H^\ddagger$ as well as log k would give an accurate linear correlation with σ.

Reaction series may be postulated where $\delta_X \Delta H = 0$; such a series is *isoenthalpic*. No really accurate illustrations exist of this type of behaviour, although figs. 5.2 and 5.3 show that for the ionisation of substituted acetic acids ΔH^0 values show only a small scatter, while the variation in pK_a is dictated mainly by variation in ΔS^0. At first sight this is a peculiar result, because the variation in pK_a is explicable on the basis

TABLE 5.2 *Thermodynamic parameters for alkaline hydrolysis of ethyl benzoates* (56 % aq. acetone, 25 °C)

Substituent	k/l mol^{-1} s^{-1} (25 °C)	$\Delta H^\ddagger$/4.18 kJ mol^{-1}	$\Delta S^\ddagger$/4.18 J K^{-1} mol^{-1}
p-NH$_2$	0.0000864	16.7	−22.9
m-NH$_2$	0.00166	14.98	−22.8
p-CH$_3$	0.00114	15.16	−22.9
m-CH$_3$	0.00169	14.87	−23.2
H	0.00289	14.56	−23.1
p-NO$_2$	0.246	12.4	−21.4
m-NO$_2$	0.137	12.8	−21.4

of inductive effects, which we have noted to be of potential energy form. We shall later see how Hepler's theory provides an explanation of this.

However, the great majority of reaction sets to which the Hammett equation has been successfully applied do not even approximately fall into the isoenthalpic or isoentropic classes. Leffler & Grunwald (1963) have proposed the more general relationship that the difference in entropy changes are proportional to the differences in enthalpy changes, and thus that kinetic and solvation energies are proportional to potential energies, i.e.,

$$\delta_X \, \Delta H = \beta \, \delta_X \, \Delta S \qquad (5.21)$$

for a reaction or equilibrium set, where β is the proportionality constant. Thence

$$\delta_X \, \Delta G = \delta_X \, \Delta H - T/\beta(\delta_X \, \Delta H)$$

so that

$$\delta_X \, \Delta G = \delta_X \, \Delta H(1 - T/\beta) \qquad (5.22)$$

Notice that the isoentropic and isoenthalpic series are special cases of (5.22), where β is infinite and zero, respectively.

The constant β has the dimensions of temperature, and when the temperature at which the experiments are conducted, T, is the same as β, equation (5.22) informs us that all the reactions in the set will proceed at the same rate, or all equilibria will have the same equilibrium constant, since $\delta_X \, \Delta G$ will be zero. A representation of such a series is given later (fig. 5.8). Constant β is thus known as the isokinetic temperature, and the theory is variously called the *linear enthalpy–entropy relationship*, the *isokinetic relationship*, or the *compensation law*.

Leffler & Grunwald demonstrated that a great number of reactions and equilibria give linear ΔH, ΔS plots, and this might well appear a partial reason at least for Hammett laws. However, a little reflection shows that it stands in complete contradiction to the point made at the beginning of this chapter. There it was noted that enthalpy and entropy changes often do *not* parallel free energy changes; the assumption that a structural variation leading to an alteration in the free energy change of a reaction also leads to a proportional alteration in the enthalpy of activation is by no means always a valid one. On the other hand, the isokinetic law, equation (5.21), by proposing proportionality between ΔH and ΔS, necessarily, from equation (5.22), implies a proportionality between ΔG and ΔH.

Petersen (1964) has demonstrated how this apparent paradox may, in part at least, have arisen; a linear ΔH, ΔS plot may originate from the fact that the errors in measurement of ΔH are proportional to ΔS. This point is illustrated with hypothetical data, assembled in table 5.3, and plotted in fig. 5.7. Reaction series 1 of table 5.3 gives in fig. 5.7 exact

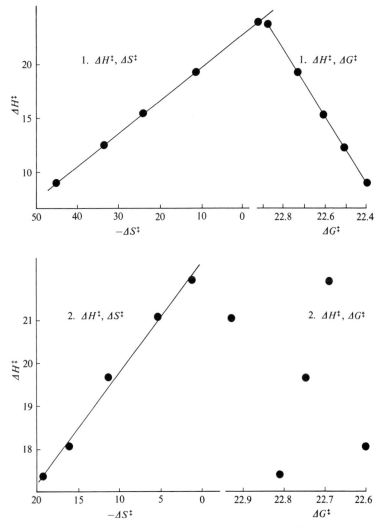

Fig. 5.7. $\Delta G^{\ddagger}$ *vs.* $\Delta H^{\ddagger}$ and $\Delta H^{\ddagger}$ *vs.* $\Delta S^{\ddagger}$ plots for hypothetical data.

linearity between any two of $\Delta G^{\ddagger}$, $\Delta H^{\ddagger}$ and $\Delta S^{\ddagger}$, but for series 2, although $\Delta H^{\ddagger}$ and $\Delta S^{\ddagger}$ yield a reasonable straight line correlation, $\Delta G^{\ddagger}$ *vs.* $\Delta H^{\ddagger}$ is a 'scattergram'. This clearly shows that a linear ΔH, ΔS plot is not sufficient to establish a valid ΔG, ΔH correlation; the former can arise from proportionality of errors in ΔH and ΔS, which are not proportional to the errors in ΔG, which derives from more accurate single temperature measurements. Thus, paradoxically, the less accurately ΔH and ΔS are measured, the greater the likelihood that they will be proportional to one another (Ritchie & Sager, 1964). Such reasoning does not, of course, deny the existence of a relationship between ΔH and ΔS, but it does show that linear plots of ΔH and ΔS are inadequate evidence for such a relationship, and thus for the precision of equation (5.22).

TABLE 5.3 *Thermodynamic parameters for hypothetical reaction data*

	$10^4 k/\text{s}^{-1}$		$\Delta G^{\ddagger}/4.18$ J mol^{-1} (25 °C)	$\Delta H^{\ddagger}/4.18$ J mol^{-1}	$\Delta S^{\ddagger}/4.18$ J K^{-1} mol^{-1}
	300 K	315 K			
Reaction series 1					
a	1.32	9.43	22.89	24.0	3.7
b	1.62	7.95	22.73	19.3	−11.5
c	1.92	6.91	22.61	15.4	−24.1
d	2.19	6.23	22.51	12.5	−33.7
e	2.56	5.46	22.40	8.9	−45.4
Reaction series 2					
a	1.2	6.8	22.93	21.1	−6.1
b	1.4	5.9	22.81	17.4	−18.2
c	1.6	8.1	22.75	19.7	−10.3
d	1.8	10.9	22.69	21.9	−2.6
e	2.0	8.9	22.60	18.1	−15.2

In any case, however, there are also a large number of sets following the Hammett equation for which neither ΔH, ΔS nor ΔG, ΔH plots are linear, three examples being given in § 5.1.

It must be noted also that the isokinetic theory predicts a correlation between ρ and $(1 - \beta/T)$ from equation (5.22). Generally, however, ρ appears to vary with $1/T$ (Ritchie & Sager, 1964; Wells, 1963), although the correlation is only approximate and more extensive data is required before a clear conclusion can be drawn (Bitterwolf, Linder & Ling, 1970).

There is, of course, abundant evidence that σ is independent of temperature, and this gives rise to yet further perplexity, for the corollary of this is that Gibbs free energy of activation must be proportional to enthalpy, if a given reaction series is to follow the Hammett equation over a range of temperatures (Hepler, 1971). Consider a reaction series composed of compounds A, B, C and D which obeys the Hammett equation at both temperature T_1 and T_2 using temperature invariant σ values. Then the plot of log k (or log K if equilibria are being considered) against $1/T$ must be as in fig. 5.8, with the consequence that $\delta_X \Delta G$ at any temperature is proportional to $\delta_X \Delta H$.

If $\delta_X \Delta G$ is not proportional to $\delta_X \Delta H$, then fig. 5.9 displays the relevant situation. Here, if the Hammett equation is followed at temperature T_1, it will not be at temperature T_2.

However, in practical situations, the range of temperature variation of ΔG that can be experimentally ascertained is only small, and predicted deviations from the equation thus difficult or impossible to detect in practice. Indeed, it is only for such limited ranges that ΔH can be assumed constant. For wide temperature variations, since the thermal

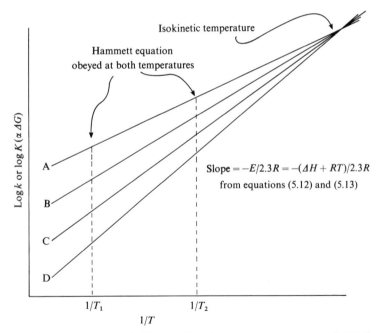

Fig. 5.8. Proportionality of ΔG and ΔH arising from temperature invariance of σ.

capacity change ΔCp $(\mathrm{d}\Delta H/\mathrm{d}T)$ is not usually zero, ΔH may vary, producing curvature in the ΔG *vs.* $1/T$ plots. This illustrates a generalisation pertinent to all the conclusions of this chapter; there is a disparity between, on the one hand, mathematical deductions based on rigorous thermodynamic principles, and on the other hand, the experimental data available for testing these deductions. Such data need to be profuse, covering a wide range of reactivity induced by medium, substituent, and temperature changes, and of a uniformly high order of accuracy, to be of any use for verification of the thermodynamic proposals. The accumulation of such results may thus not even be experimentally feasible. For example, theoretically arguing that ΔCp is not zero for a given set of reactions is one thing; experimental establishment of accurate ΔCp values for the reactions by measurement of ΔH with sufficiently high exactitude to disclose its exact variation with temperature is quite another.

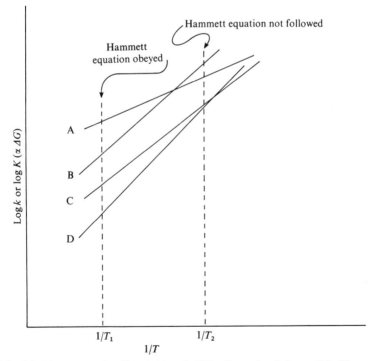

Fig. 5.9. Non-proportionality of ΔG and ΔH leading to breakdown of the Hammett equation at temperatures other than T_1 when temperature invariant σ values are used.

5.5. Internal and external contributions to ΔH and ΔS. The apparent defects in the isokinetic theory have been countered, at least partially, by the more detailed analysis due to Hepler (Larson & Hepler, 1969). In this, contributions to ΔH and ΔS are considered in terms of external or environmental and internal factors, given by ΔH_{env}, ΔS_{env} and ΔH_{int}, ΔS_{int}. Hepler considers that environmental effects arise from solute–solvent interactions, while internal effects originate in enthalpy and entropy differences between the molecules on either side of an equilibrium or between reactant molecules and the activated complex. ΔS_{int} is taken to be zero for symmetrical reactions of type [5.2] and [5.3] where intramolecular hydrogen bonding or steric effects often associated with o-substitution are absent, so that

$$\delta_X \Delta S = \delta_X \Delta S_{env} \qquad (5.23)$$

That $\delta_X \Delta S_{int}$ may be taken as zero is illustrated nicely by work of Taft (Love, Cohen & Taft, 1968), who has studied the symmetrical reaction

$$XCH_2\overset{+}{N}H_2\!-\!\overset{-}{B}(CH_3)_3 + CH_3CH_2NH_2 \rightleftharpoons$$

$$XCH_2NH_2 + CH_3CH_2\overset{+}{N}H_2\!-\!\overset{-}{B}(CH_3)_3 \qquad [5.4]$$

in the gas phase at 54 °C, where obviously interactions of solute–solvent type are absent. The relevant thermodynamic parameters are set out in table 5.4.

The $\delta_X \Delta S^0$ terms are zero, within experimental error, with the exception of those of the methoxy-substituted derivatives; here an intramolecular interaction of the form of (77) can be distinguished.

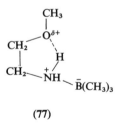

(77)

Essentially zero entropy changes are also to be found in the vapour phase pyrolysis of 1-arylethyl acetates (Smith & Kelly, 1971).

TABLE 5.4 *Dissociation of gaseous trimethylboron adducts of primary amines* (54 °C)

X	$\delta_X \Delta G^0/4.18$ kJ mol^{-1}	$\delta_X \Delta H^0/4.18$ kJ mol^{-1}	$\delta_X \Delta S^0/4.18$ J K^{-1} mol^{-1}
C_2H_5	(0.00)	(0.0)	(0.0)
CH_3	−0.07	−0.1	0.0
CH_3CH_2	+0.21	+0.3	0.2
$CH_3O(CH_2)_3$	+1.09	+1.8	+2.2
$CH_3O(CH_2)_2$	+0.41	+1.3	+2.7
$F(CH_2)_2$	−1.29	−1.3	0.0
F_2CHCH_2	−2.83	−2.8	0.0
F_3CCH_2	−4.00	−4.0	0.0

Since $\delta_X \Delta H_{env}$ and $\delta_X \Delta S_{env}$ arise from the same cause, solvent–solute interactions, it is not unreasonable to assume that $\delta_X \Delta H_{env}$ is proportional to $\delta_X \Delta S_{env}$, i.e.,

$$\delta_X \Delta H_{env} = \beta_{env} \delta_X \Delta S_{env} = \beta_{env} \delta_X \Delta S \qquad (5.24)$$

β_{env} is reasoned to be of a similar magnitude to the temperature of the experiments, and to be the same for similar reaction types. Thus it is 280 K for the ionisation of both aliphatic and aromatic carboxylic acids and phenols.
We may then write

$$\delta_X \Delta H = \delta_X \Delta H_{int} + \beta_{env} \delta_X \Delta S$$

whence, from equation (5.20),

$$\delta_X \Delta G/\delta_X \Delta H_{int} = 1 + (\beta_{env} - T)\delta_X \Delta S/\delta_X \Delta H_{int} \qquad (5.25)$$

Since $\beta_{env} \cong T$, and $\delta_X \Delta S/\delta_X \Delta H_{int}$ is considerably less than unity:

$$\delta_X \Delta G \cong \delta_X \Delta H_{int} \qquad (5.26)$$

We have thus arrived at an approximate verification of the conclusion which was noted at the start of this chapter; substituent effects on the free energy parameters, logarithmic rate or dissociation constants, can often be simply explained in terms of electronic factors, which are measured by ΔH_{int}, whereas overall ΔH and ΔS values display a variation, because of environmental contributions, often much more complex and therefore difficult to rationalise.

It may further be noted that for a reaction obeying the Hammett equation

$$\rho\sigma = -\delta_X \, \Delta G/2.3 \, RT \simeq -\delta_X \, \Delta H_{int}/2.3 \, RT \qquad (5.27)$$

i.e.,

$$\rho \propto 1/T \qquad (5.28)$$

in agreement with approximate experimental observation. One obvious criticism of (5.27) is that it does not cater for the variation of ρ with solvent, although Larson & Hepler (1969) have indicated how approximate provision may be made for this. It certainly does not seem unlikely that future interpretation of enthalpy–entropy data for reaction sets will be along the lines of this theory, sketched in briefly here.

5.6. The Hammond postulate. It is appropriate at this stage to consider a general theorem relating to transition state structures. This has been referred to variously as the Ogg–Polanyi–Hammond postulate (Pietra, 1969), the Bell–Evans–Polanyi principle (Dewar, 1969), or the Hammond postulate (Hammond, 1955). Here we shall use the latter name, the choice dictated solely by considerations of brevity.

The postulate deals with the association between the rates and transition state structures of mechanistically analogous reactions. Imagine the reactions between reactant molecules X and Y_1 and Y_2, which combine by the same mechanism in the rate-determining step to form XY_1 and XY_2, passing through transition states T_1 and T_2:

$$X + Y_1 \xrightarrow{\ T_1\ } XY_1 \qquad [5.5]$$

$$X + Y_2 \xrightarrow{\ T_2\ } XY_2 \qquad [5.6]$$

Thus Y_1 and Y_2 might be two esters and X represent OH^-, so that XY_1 and XY_2 are the tetrahedral intermediates and the reactions are of the form [3.2]; or X could be benzene, Y_1 and Y_2 different electrophiles and XY_1 and XY_2 Wheland intermediates as in [2.3]. If transition state T_1 is closely allied to the ground state $X + Y_1$, in that bonding between X and Y_1 is only weak, it can be argued that the potential energy difference between the two states, which will be the activation energy E, should be only small. Alternatively, if T_2 resembles XY_2, so that bonding is strong and the reaction has travelled far along the reaction co-ordinate, E will be larger. The potential energy diagram

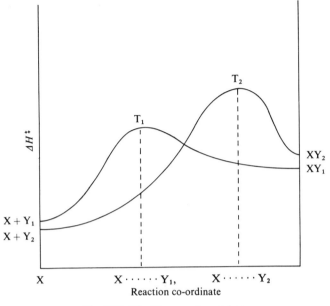

Fig. 5.10. The Hammond postulate.

thus surmised is shown in fig. 5.10. Consequently, assuming $\Delta G^{\ddagger}$ to be proportional to $\Delta H^{\ddagger}$ and therefore E, reaction [5.5] will proceed faster than [5.6].

The corollary to this is the *selectivity principle*; the more reactive the reagent X, the less is its selectivity, and the smaller will become the variations in rate as reagent Y is varied. In the limit, as reagent X becomes infinitely reactive, its rate of reaction with all molecules of type Y will become infinitely fast, and the transition state T of the reactions will be represented by the ground state of such Y molecules.

It must be noticed that the argument has been derived in potential energy terms, and the validity of the isokinetic relationship assumed. However, as we have seen previously, proportionality between enthalpy and free energy is by no means invariably the case. In many instances, probably at least partially for this reason, the Hammond postulate apparently breaks down, examination showing that it is often inadequate as a general postulate, *assuming that it is correct to regard reaction constants ρ as an accurate indication of transition state character* (as argued on §1.3 and in many examples subsequently). Let us look at some of the evidence.

Although it has often been assumed that the less selective an electrophile the faster its reaction with benzene, absolute rate measurements, where they do exist, often do not bear this out. The Hammond postulate is certainly not well substantiated for electrophilic substitution, and it is certain that there is little correlation between the ρ values for such substitutions and the rate of attack of the various electrophiles on benzene. Thus Dubois, Alcais & Rothenburg (1968) have shown that while replacing acetic acid with trifluoroacetic acid as solvent accelerates bromination of benzene and methylbenzenes by a factor of approximately 10^7, the ρ value changes from -10 in acetic acid to -11 in trifluoroacetic acid. Clearly the transition states are similar; if anything, the transition state for the faster reaction is nearer the Wheland intermediate.

The same variation in solvent also increases the rate of electrophilic mercuration (Brown & Wirkkala, 1966), this time by a factor of 10^6. However, the ρ values, -4.0 in acetic acid and -5.7 in trifluoroacetic acid, signify a more reactant-like transition state in the former case.

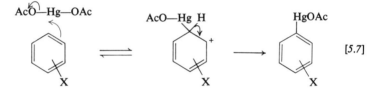

$$[5.7]$$

A possible complication here is that a large isotope effect is observed for the attack of this electrophile, so that the precise mechanism of the reaction is in doubt (Hoggett, Moodie, Penton & Schofield, 1971); change of solvent may indeed change the rate-determining step.

Furthermore, failure of the Hammond postulate induced by solvent variation or change in reagent is perhaps not unexpected, because changes in solvation interactions at the reaction site will manifest themselves in irregular enthalpy and entropy terms. But just as we saw that, even where the molecular variations were solely substituent changes well removed from steric interactions with the reaction site, proportionality between $\Delta G^{\ddagger}$ and $\Delta H^{\ddagger}$ was not observed, so even for this same apparently simplified situation, the Hammond postulate may fail.

A clear example comes from the work of Norman (Knowles, Norman & Prosser, 1961), who investigated the displacement of chloride ions from 1-chloro-2,4-dinitrobenzene and 1,4-dichloro-2-nitrobenzene by *m*- and *p*-substituted phenoxide ions, both reactions being carried out in 80%

aqueous dioxan at 65 °C. The dinitro compound reacted some 10^5 times faster than the mononitro compound, but the ρ values for the two reactions were both -1.8, indicating a close similarity of transition state structure. Another illuminating example comes from the work of Kirsch, Clewell & Simon (1968). The effect of substituents, in both the acyl and aryl moieties of phenyl benzoates, on the rate of alkaline hydrolysis was studied. The rate constant variation range was greater than 10^5, but the ρ values, shown in table 5.5, indicated a constant transition state.

Indeed, careful thought leads to the conclusion that *the reasoning commonly attached to ρ values defined by the Hammett equation is completely contrary to the Hammond postulate*, because while the former attempts to determine the nature of the transition state of a given reaction by changing the rate through substituent variation, the latter

TABLE 5.5 ρ *values for alkaline hydrolysis of phenyl benzoates* (33 % *aq. acetonitrile* 25 °C)

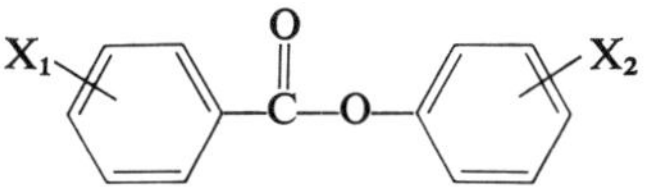

Substituent X_1 constant, ρ determined by varying X_2

X_1	Relative rate ($X_2 = H$)	ρ
$p\text{-N(CH}_3)_2$	0.02	1.20
$p\text{-CH}_3$	0.42	1.28
$p\text{-H}$	1	1.28
$p\text{-Cl}$	3.1	1.18
$p\text{-NO}_2$	35	1.25

Substituent X_2 constant, ρ determined by varying X_1

X_2	Relative rate ($X_1 = H$)	ρ
$p\text{-CH}_3$	0.60	2.03
$p\text{-H}$	1	1.98
$p\text{-Cl}$	2.2	2.04
$m\text{-NO}_2$	8.3	2.04
$p\text{-NO}_2$	14	2.01

tells us that the character of the transition state will be altered by the very process of changing the rate.

Furthermore, the Hammett equation is not alone in this assumption; an immutable transition state is generally assumed in MO calculations on the effect of structure on reactivity, not only in the simple methods described in chapter 4, but also in the more sophisticated ones, as well as in the Brønsted equation, which equates the very wide rate variations in acid- and base-catalysed reactions with the pK_a's of these catalysts.

This puzzling discrepancy between the Hammond postulate and the Hammett equation would be resolved if the changes of free energy of activation induced by change of substituents in application of the latter was small compared with the average magnitude of the total free energy of activation. This would correspond, in the analogy given at the beginning of § 1.4, to the effect of the low conductance of a voltmeter, which enables the detection and measurement of potential by the instrument with minimum perturbance of that potential and thus of the electronic characteristics of the circuit.

Unfortunately, the analogy breaks down at this point. From equation (5.9), the free energy change will be given by

$$\delta_X \, \varDelta G^{\ddagger} = 2.3 \, RT \, \rho \, \varDelta \sigma \qquad (5.29)$$

Application of equation (5.29) to a reaction at room temperature, of ρ value 2, established by substituent variation from p-OCH$_3$ to p-NO$_2$ so that $\varDelta\sigma$ is 1.05, yields a value of $\delta_X \, \varDelta G$ of 11 kJ mol^{-1}. This is to be compared with a typical $\varDelta G^{\ddagger}$ value of about 60 or 70 kJ mol^{-1}, of which it is certainly not an insignificant fraction. In many instances, of course, ρ is greater than 2, and $\varDelta\sigma$ is greater than 1.05, so that the total range of reactivities accurately correlated is frequently 10^6 or more; there is thus no evidence for deviance from the Hammett equation when $\delta_X \, \varDelta G^{\ddagger}$ forms a very much higher contribution to $\varDelta G^{\ddagger}$. Indeed, it is quite extraordinary how the same set of σ values will correlate the whole spectrum of reactivity changes induced by substituents, between on the one hand those involving such high energy variations as discussed here, and on the other the minute energy alterations involved for example in NMR chemical shift experiments.

In summary, the following two principles appear to be implicated in the accepted theoretical reasoning behind the Hammett equation.

1. The alteration of the rate of a given reaction by attachment of substituents, so that steric interactions between them and the reaction

site are negligible, depends on the differential stabilisation of the ground
and transition states, but involves only small changes in the positioning
of the transition state along the reaction co-ordinate, so that variation
of the molecular co-ordinates of the transition state reaction site are
minimal.

2. The Hammond postulate is incorrect; there is no general relationship between rates of reactions of similar mechanistic characteristics,
and the position of the transition state along the reaction co-ordinate.

If, alternatively, the Hammond postulate is upheld, then a necessary
consequence is that ρ is not in anyway a realistic indication of the nature
of the transition state. The Hammond postulate would appear to indicate
that Hammett plots, particularly those covering a large reactivity range
(those with a large ρ value coupled with study of a large σ range) should
be curved. Generally, however, such curvature cannot be detected even
for such optimum cases (e.g. figs. 2.6, 2.7 and 4.3). Where curvature does
occur (e.g. figs. 2.13, 2.15 and 2.18), it appears convincingly explicable
in alternative form, namely as due to mechanistic changeover, or the
onset of diffusion control (fig. 2.9).

However, it is essential to note that the idea of ρ as indicative of
transition state structure stems from chemical intuition and qualitative
reasoning; the Hammett equation itself makes no reference to the
positioning of transition states along the reaction co-ordinate. Like the
term steric hindrance, that of transition state has become commonplace
by repetition, but since it represents the most unstable species encountered in the passage of reactants to products, its complete experimental
and theoretical definition is, as for steric hindrance, bound to be impossible. While ρ values can be 'explained' in terms of a constant
transition state, a more sophisticated treatment involving the notion of
a variable transition state is probably not impossible. It is certain that
such a treatment would also have to take into account not only the
variation in transition state geometry and energy but also the equivalent
variations in the reagents and primary products.

Perhaps we should finish on this cautionary note. The theoretical
interpretation of the Hammett equation has been built up from circum-
stantial evidence rather than by rigorous proof, although the number
and variety of successful correlations that have emanated from it are
remarkable, considering the simplicity of the approach. It has aptly
been described (Bolton & Hepler, 1971) as more general than 'it ought
to be'. However, *if the validity of the approach is accepted, and ρ values*

agreed to be a legitimate measure of transition state structure in a kinetic process, then it is important to use the equation consistently. For example, it is quite incorrect to construe it as an indication of difference in transition state structure for the same reaction in two different nuclei in which there is no steric interaction between the site of reaction and the components of either nuclei. This is a consequence of the general application of equations of the form of (4.6) to (4.9).

An example makes this clear. Butler & Hendry (1970) have measured the rates of chlorination, in acetic acid, of the 2-position in 5-substituted thiophenes. Reasonable correlation between the log rate constants and σ^+_p were found, giving a ρ value of -6.5. Since benzene reacts some 10^7 times more slowly than thiophene, and for substituted benzenes the ρ value is -10.0, we are very tempted to regard this as an indication of the applicability of the Hammond postulate, in that the more reactant-like transition state, bearing the smaller positive charge, is found in the faster thiophene reaction. But such a conclusion would be completely incompatible with use of the Hammett equation; the interpretation of the two ρ values has involved the assumption that the transition state is *not* altered by substituents, and its character is essentially independent of reaction rate. This becomes apparent when it is noted that, although benzene reacts much more slowly than thiophene, p-chlorination of methoxybenzene (table 2.5), which is one of the reactions used to define the value of -10.0, reacts considerably *faster* than α-carboethoxythiophene, the rate for which fits the reported ρ value of -6.5. The correct explanation for the different ρ values is thus experimental error and inexactitudes, or steric or proximity interactions between the sulphur atom with the reaction site. It is significant that Butler (1970) reports that the ρ values for reactions of the two systems are generally very similar, as noted in § 4.4; such a result is required by the success of the extended selectivity relationship when applied to this and other five-membered ring systems, which clearly demonstrates an immutable σ value for the ring nitrogen, and steric interactions to be commonly absent, and ρ values equivalent to those for benzenoid reactions.

To summarise, one can clearly discern, as a general conclusion to the issues discussed in these pages, that future progress in our recognition of the theoretical implications of the Hammett equation lies in the quest for a fundamental interpretation of its unique and often disconcerting simplicity, and not in the imposition of an artificial complexity on a treatment which is quite plainly reluctant to accept it.

5.7. Problems

26. Rate constants and Arrhenius activation energies for reaction of substituted fluorobenzenes with methoxide ions in methanol at 100 °C are set out below:

Substituents	$10^9 \, k/\text{l mol}^{-1} \text{s}^{-1}$	$E/4.18 \text{ kJ mol}^{-1}$	σ^-
H	6.88	36.4	
3-CF$_3$	3.32×10^3	26.7	0.43
3,5-(CF$_3$)$_2$	1.17×10^6	22.6	
4-CF$_3$	4.05×10^4	22.5	0.74
3-SO$_2$CH$_3$	5.38×10^4	26.5	0.60
4-SO$_2$CH$_3$	5.13×10^6	24.2	1.05
3-NO$_2$	7.17×10^4	28.9	0.71
4-NO$_2$	1.65×10^8	20.1	1.27
3,5-(NO$_2$)$_2$	4.06×10^8	21.7	
3-CF$_3$, 5-NO$_2$	3.23×10^7	22.6	
3-SO$_2$CH$_3$, 5-NO$_2$	1.78×10^8	22.4	

(Hirst & Una, 1971)

Examine the relevance of the Hammett equation to the reaction, comment on the ρ value obtained and the correlation of $\Delta G^{\ddagger}$ with $\Delta H^{\ddagger}$. Calculate k values for 60 and 140 °C and examine the correlation with the Hammett equation at this temperature. Consider the applicability of the Hammond postulate to this reaction, and hence comment on the fact that the ρ values for the reaction of substituted 1-chloro-2-nitrobenzenes and chlorobenzenes with methoxide ions in methanol at 50 °C are 3.6 and 8.5 respectively.

27. The following table shows data for the TiCl$_4$ catalysed reaction of XC$_6$H$_4$CH$_2$Cl at 30 °C with benzene and toluene (Olah, 1971).

X	$k_{\text{toluene}}/k_{\text{benzene}}$	% p-substitution in toluene
NO$_2$	2.5	34.2
H	6.3	55.2
CH$_3$	29.0	66.5
OCH$_3$	97.0	69.9

Calculate the ρ values for the reactions. Assuming that benzene reacts 100 times faster when X = NO$_2$ than when X = OCH$_3$, calculate the σ^+ value for the substituted benzene for which reactivity with the two electrophiles becomes equal, and the selectivity–reactivity (Hammond) relationship breaks down.

Repeat the calculations to establish ρ values for the $AlCl_3$ catalysed reactions of XC_6H_4COCl (CH_3NO_2, 25 °C) given below, and again establish a σ^+ value for which reactivity becomes equal, assuming a rate difference of $X = NO_2 > X = OCH_3$ of 100 for benzene.

X	$k_{toluene}/k_{benzene}$	% p-substitution in toluene
NO_2	52	89.3
OCH_3	233	83.6

Problem discussion

1. Log relative rates *vs.* σ yield a good straight line plot (fig. A):

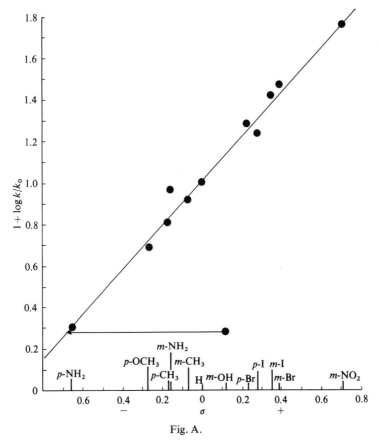

Fig. A.

The ρ value is 1.07, positive since rate of attack of the nucleophile OH⁻ is enhanced by electron-withdrawing groups. Deviations from the line, with the exception of the OH point, typify the degree of error to be generally expected. The extent of the OH deviation however clearly indicates a marked change in electronic character of the group; in alkaline media the substituent will be ionised forming O^-, for which the effective σ_m value of -0.69 indicates the expected strong $+I$ effect.

2. $NHCOCH_3 : \sigma_I = 0.21$, $\sigma_R = -0.21$.

$$\overset{\displaystyle \text{O}}{\underset{}{\rightarrow}} \; \overset{\displaystyle \parallel}{\rightarrow \overset{..}{N}H \rightarrow C - CH_3}$$

The inductive electron withdrawal by N is enhanced by attachment of $COCH_3$. Conjugation between the nitrogen lone pair and the carbonyl group reduces the extent of alternative conjugation with the benzene ring, as present, for example, in aniline.

$C_6H_5 : \sigma_I = 0.06$, $\sigma_R = -0.07$.
The group appears to withdraw electrons by induction and donate them by resonance, but both effects are very small. This is in line with the basic idea of the Hammett equation, which assumes that the perturbation of the substituted nucleus by the reacting side chain is minimal. If this assumption were precisely correct, σ_m and σ_p should be zero for C_6H_5.

$CF_3 : \sigma_I = 0.43$, $\sigma_R = 0.11$
This substituent shows the expected large positive σ_I value due to the three strongly electronegative fluorine atoms, together with a small resonance effect due to fluorine hyperconjugation.

$$-CF_3 \leftrightarrow -\overset{+}{C}F_2F^-$$

$CH{=}CH{-}NO_2 : \sigma_I = 0.32$, $\sigma_R = -0.06$
The inductive effect of the group can be calculated as that of NO_2 diminished by passage through two sp^2 carbons, $(\sqrt{0.47})^2 \times 0.71$, which yields 0.33, in good agreement with the observed value. The negative value of σ_R shows that the expected resonance

$$\overset{\downarrow}{\underset{-}{C}H{=}CH{-}\overset{+}{N}\overset{\displaystyle \nwarrow O}{\underset{\displaystyle O^-}{\nearrow}}}$$

is not in fact present; rather the π electrons are conjugated with the benzene ring, so the total electronic effect of the group is similar to that of halogen.

$C_3H_5 : \sigma_I = -0.07$, $\sigma_R = -0.14$
This cyclic substituent has a much greater resonance-donating capacity than alkyl groups; its unique conjugative ability is confirmed by

spectroscopic studies, and probably arises from sp^2 hybrid character of the C atoms.

$SOCH_3 : \sigma_I = 0.52, \sigma_R = -0.03$; $SO_2CH_3, \sigma_I = 0.60, \sigma_R = 0.12$

The interaction of the first substituent with the benzene ring is almost entirely inductive; the negative σ_R value is too small to be significant. σ_m/σ_p is 1.07, in agreement with the equivalent ratio for $\overset{+}{N}(CH_3)_3$. The introduction of a second electronegative oxygen atom introduces additional electron withdrawal, partially by induction and predominantly by resonance:

$OCF_3 : \sigma_I = 0.40, \sigma_R = -0.05$

The electronegative CF_3 group increases the inductive withdrawing property of OCF_3 compared with OCH_3. The resonance donation of the latter group is decreased in the former case by the alternative fluorine hyperconjugation:

$$-\overset{..}{O}-CF_2-F$$

3. The σ_m values of 0.36 and 0.35 show good agreement, well within experimental error.

CHO would be expected to have both a positive σ_I and σ_R value:

$$\rightarrow CH = O$$

and thus the σ_p value of 0.43 is predicted to be the correct one.

The anomalous value of 0.22 may be due to a base-catalysed condensation between the formyl group and acetone employed as solvent:

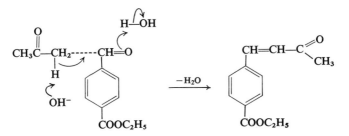

This reaction will be faster in the *p*-substituted compound owing to the increased electron withdrawal by *p*-COOC$_2$H$_5$ ($\sigma = 0.45$), compared with *m*-COOC$_2$H$_5$ ($\sigma = 0.37$), accelerating the rate anion attack occurs on the carbonyl of the formyl group. Therefore the σ_p value may refer to CH=CH—COCH$_3$, comparable with that of 0.26 for the electronically analogous group CH=CH—NO$_2$.

4. There is apparently no significant resonance electron withdrawal by the picryl substituent, because the ratio σ_m/σ_p is 1.05, in good agreement with the value for other substituents with a predominant $+I$ effect. The absence of conjugation indicates that there is considerable steric hindrance between *o*-H and *o*-NO$_2$, so that the rings are orthogonal:

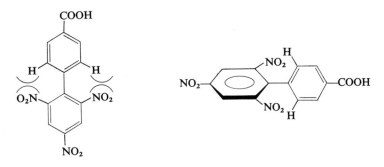

5. If the additivity principle is assumed for these compounds, the effective $\sum \sigma$ values will be $(-0.07 -0.07 +\sigma_p)$. $\sum \sigma$ values are thus:

N(CH$_3$)$_2$	-0.97	Cl	0.09	CN	0.52	NO$_2$	0.64
NH$_2$	-0.80	Br	0.09	COOCH$_3$	0.31		

where COOCH$_3$ has been assumed to have the same σ_p value as COOC$_2$H$_5$.

The line drawn in fig. B is that defined by the monosubstituted benzoic acids, slope 1.60, passing through (5.71, 0). Groups *p*-Cl, *p*-Br, *p*-CN and *p*-NH$_2$ all fall on or very near to this line, demonstrating the absence of twisting by the flanking CH$_3$ substituents. This is a reasonable conclusion because two groups are monatomic, the triple bond of the third (CN) is in the same line as the bond connecting it to the benzene ring, and the fourth (NH$_2$) has only two protons joined to N.

The twisting of COOCH$_3$, NO$_2$ and N(CH$_3$)$_2$ from the plane of the ring must be very considerable and accordingly their resonance interaction is severely reduced. If we assume that it is completely eliminated,

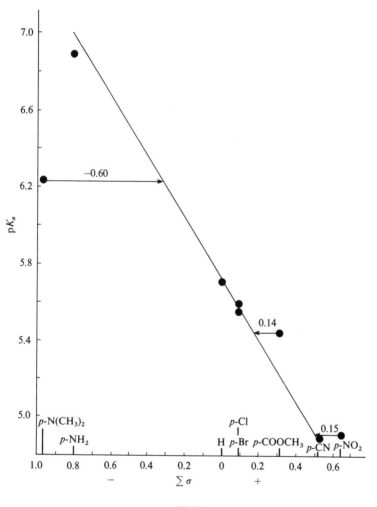

Fig. B.

then we have the following σ_I and σ_R values, to be compared with those in brackets derived from equations (1.12) and (1.13):

$$p\text{-NO}_2: \quad \sigma_I = 0.63 \ (0.71), \qquad \sigma_R = 0.15 \ (0.07)$$

$$p\text{-COOCH}_3: \quad \sigma_I = 0.31 \ (0.37), \qquad \sigma_R = 0.14 \ (0.08)$$

$$p\text{-N(CH}_3)_2: \quad \sigma_I = -0.23, \qquad \sigma_R = -0.60$$

The negative value for the estimate of σ_I of $N(CH_3)_2$ shows that some resonance interaction is still present.

The additivity of the Hammett equation is thus very good, provided steric interactions are absent. A proximity effect leading to breakdown of additivity, other than the twisting of groups, is hydrogen bonding between one substituent and a second.

6. Assuming additivity of substituent effects as in the previous question, the pK_a value of 3,5-dimethyl-4-nitrophenol is given by the following equation, using σ^- for NO_2:

$$-pK_a + 9.99 = 2.11 \times (-0.07 - 0.07 + 1.27)$$

therefore

$$pK_a = 7.61$$

The predicted pK_a indicates that the molecule should be $0.64\ pK_a$ units more acidic than observed. As reasoned in the previous question, the flanking CH_3's twist NO_2 out of the plane of the ring, reducing its resonance stabilisation of the acid anion and thus its effective σ^- value. A similar calculation for 3,5-dimethyl-4-cyanophenol gives a predicted pK_a value of 8.43, only 0.22 units away from the observed value. In this case the agreement is better, because as we noted previously, CN is a linear group.

7. The log relative rates yield a reasonable straight line when plotted against σ^+, of slope (ρ) -4.6 (fig. C). The most likely mechanism is thus aromatic electrophilic substitution involving cleavage of the aryl–Si bond by an attacking proton:

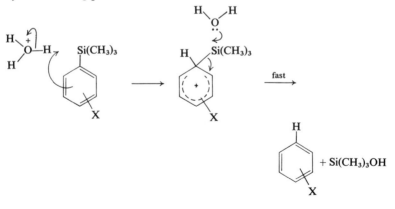

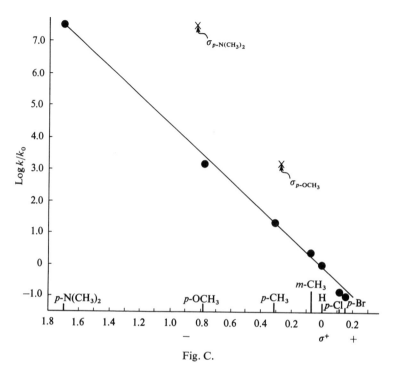

Fig. C.

The rate for $N(CH_3)_2$ was estimated by experiments in weaker acid where the free base concentration could be ascertained.

8. It is doubtful if a fully-fledged acyl nitrene is ever developed in the Lossen rearrangement. Probably the electronic shifts shown are synchronous, with the cleavage of the N—O bond running somewhat ahead of the others:

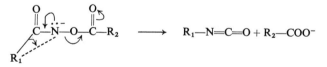

The ability of R_1 to attack the electron deficient N atom with its electron pair will be enhanced by electron donors, so here ρ will be negative – this part of the reaction is essentially intramolecular aromatic electrophilic substitution. The formation of R_2COO^- will display a similar dependence on substituents as the ionisation of benzoic acids.

9. Clearly the carbonyl group must be implicated in the reaction, or the substituents X would be so far removed from the reaction site that their effects would be negligible. A scheme which accommodates this and the labelling evidence is:

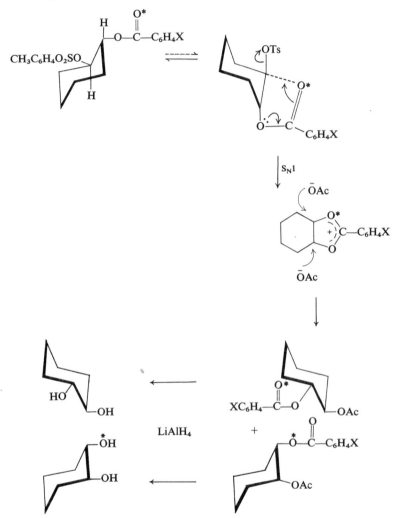

A fit with σ rather than σ^+ indicates that the positive charge in the S_N1 intermediate is fully stabilised by the two adjacent oxygen atoms, and does not therefore require additional electron donation from substituents such as OCH_3:

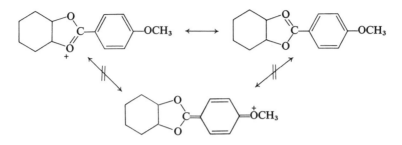

10. In neutral solution, the nucleophile is water and a tetrahedral intermediate is formed:

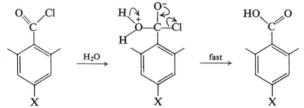

The rate of nucleophilic attack will be sponsored by electron-withdrawing groups; hence ρ is positive.

In acidic solution, the concentration of nucleophile H_2O is reduced by formation of H_3O^+, and the mechanism is:

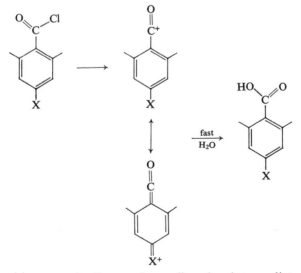

The transition state leading to the acylium ion intermediate will be stabilised by electron donors in a σ^+ type correlation.

11. The change from good to poor leaving groups, indicated by the relative reactions rates, entails the transition state carrying progressively more negative charge on C_β, progressing towards an El_{cb} type mechanism, and thus giving an increasingly positive ρ value.

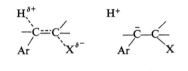

Good leaving group   Poor leaving group

Because the ρ value indicates accumulation of negative charge on C_β, the correlation will be with σ^-. A maximum isotope effect, k_H/k_D, occurs when the hydrogen bond is exactly half broken in the transition state; the values given in the final column thus indicate that in β-aryl ethyl iodides the bond is at least half broken, and progressively more than half-broken in the subsequent members of the series.

12.

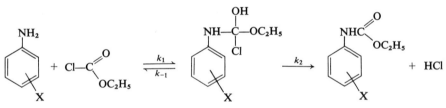

This reaction is a specific example of the generalised scheme equation (2.2). When electron-withdrawing substituents are present, the amines are less nucleophilic, and $k_{obs} = k_1$, equation (2.4). Electron donor groups make the amines more nucleophilic, so that $k_{obs} = k_1 k_2/k_{-1}$, equation (2.6). The very much increased negative ρ value, -1.57 to -5.56, suggests participation of the N electron pair in the second stage of the reaction:

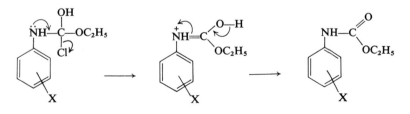

13. (1) The absence of any significant substituent effect suggests the $A_{Ac}2$ mechanism, in which substituent effect on the initial protonation almost exactly balances that on the formation of the tetrahedral intermediate:

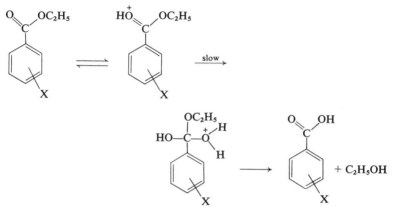

(2) In this strongly acid medium, the esters will be completely protonated, and since there is little water present, the mechanism will be of $A_{Ac}1$ form, involving acylium ion formation assisted by electron donor substituents:

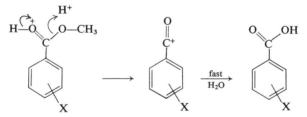

It is perhaps rather surprising that the correlation is with σ^+ values rather than σ values, because the through conjugation in the acylium ion could be matched by through conjugation in the protonated ester:

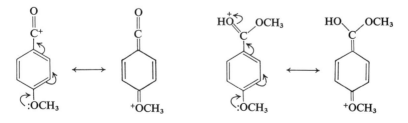

However, the latter conjugation is limited by the alternative resonance:

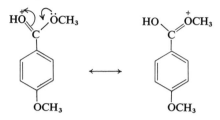

This explanation is corroborated by the correlation of the pK_a's of amides, where analogous resonance in the protonation site can occur, with σ, whereas the pK_a's of benzaldehydes, in which such resonance is absent, is with σ^+.

(3) In this case, the stability of the *iso*propyl carbonium ion encourages the $A_{Al}1$ (S_N1) rather than the $A_{Ac}1$ mechanism as the predominant route to products, the formation of the carboxylic acid from the protonated ester being favoured by electron-withdrawing groups:

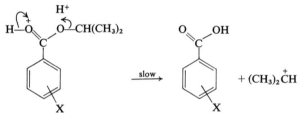

(4) Here a mechanistic transition has occurred owing to the intermediate stability of $C_2H_5^+$ between CH_3^+ and $(CH_3)_2\overset{+}{C}H$. The $A_{Ac}1$ mechanism is favoured by electron donors and $A_{Al}1$ by electron acceptors.

(5) Formation of the aryl stabilised carbonium ion ($A_{Al}1$ or S_N1 mechanism) is favoured over the alternative $A_{Ac}2$ mechanism, even in the presence of a high water concentration:

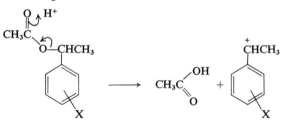

14. Substitution of the figures into equation (3.3) yields values of E_s as shown below:

Acid	E_s	Ester	E_s
CH_3CH_2COOH	−0.06	$CH_3CH_2COOC_2H_5$	−0.08
$(CH_3)_3CCOOH$	−1.60	$(CH_3)_3CCOOC_2H_5$	−1.54
$ClCH_2COOH$	−0.17		
$BrCH_2COOH$	−0.26		
ICH_2COOH	−0.33		

Comparison of the figures in the two columns together with those in tables 3.2 and 3.4, the latter being averaged values for all the data available, show that there is reasonable agreement in the E_s parameters and that they are therefore sensibly independent of reaction (esterification or hydrolysis) and solvent type.

15. Substituting the figures given into equation (3.6), using values of E_s of −0.07 and −1.6 for ethyl propionate and trimethylacetate, yields the values of σ^* given below:

	σ^*	
	aq. acetone hydrolysis	aq. alcohol hydrolysis
CH_2CH_3	0.14	0.09
$C(CH_3)_3$	0.33	0.34

The agreement of these figures with the averaged values in tables 3.3 and 3.4 confirms the approximate solvent independency of σ^* values. It is worth emphasising again that this independency does not extend to pure water.

16 and 17. Both of these questions include data for which the Yukawa–Tsuno equation, rather than the simple Hammett equation, might have been considered appropriate. The reason for this in the case of semicarbazone formation has already been discussed in § 2.9, and indeed an r value of 0.40 included in table 3.8, has been reported. However, an excellent fit with σ values is obtained (see fig. D; see also fig. 2.17).

The large substituent effect in the S_NAr reaction serves to indicate that k_1 is the observed rate constant in the scheme;

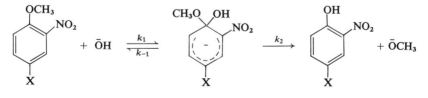

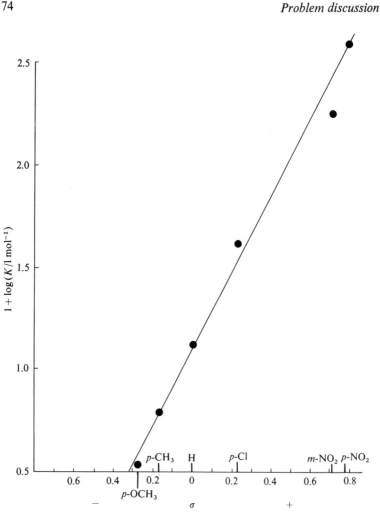

Fig. D.

If the observed rate constant was $k_1 k_2/k_{-1}$, the ρ value governing the initial equilibrium k_1/k_{-1} would be similar in magnitude but opposite in sign to the ρ value for the second step, and thus ρ_{obs} would be small.

Values between σ and σ^- might be expected for the correlation of CF_3 and NO_2, since the transition state leading to the formation of the

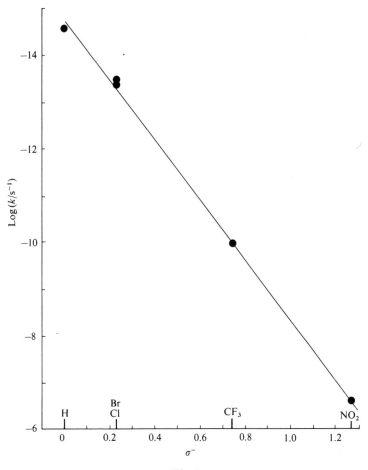

Fig. E.

Meisenheimer complex involves less than complete delocalisation of negative charge into the ring:

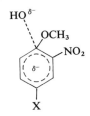

However, the ρ value of 6.0 is obtained using σ^- values (see fig. E). (In § 3.3, dimethyl sulphoxide was shown to have unique solvation properties. One of these is its inability to stabilise negative charge. A consequence of this is that when this reaction is conducted in dimethyl sulphoxide the resonance withdrawal of CF_3 and NO_2, leading to separation of negative charge, is suppressed, and the correlation is with σ.) These two examples illustrate the practical effectiveness of the $\sigma, \sigma^+, \sigma^-$ trichotomy, however oversimplified it may appear theoretically.

18. Log k is correlated best with σ^+, giving a ρ value of -0.7. The rate-determining step is consequently predicted to be of the following form:

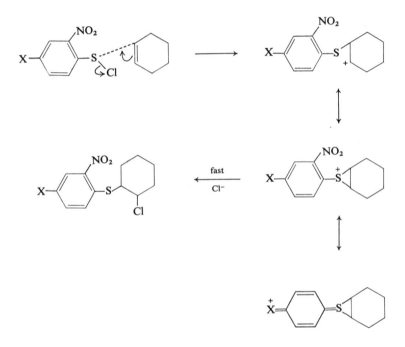

Plotting the log rate constants for substituents in the cyclohexane ring *vs.* σ_I, taking the latter values from table 3.1, apart from that for $C(CH_3)_3$, which comes from table 3.3, a linear correlation is obtained (see fig. F). This is reasonable, because the effect of X and X' can only be relayed by induction to the reaction site.

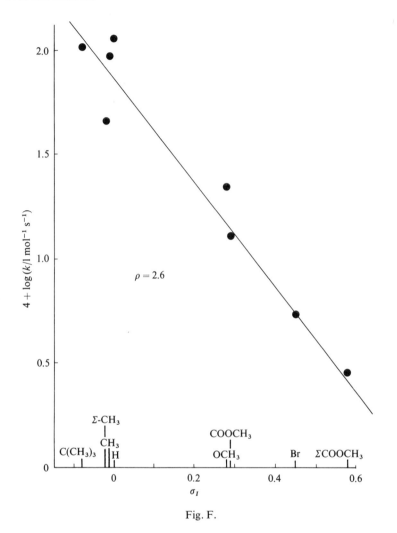

Fig. F.

19. In 77 % acid there is considerable scatter of the points, indicating steric hindrance in the transition state between substituents and reaction centre, for which the obvious mechanism is $A_{Ac}2$. This is supported by the (very) approximate ρ value of 0.7. In the stronger acid, two linear correlations are produced, a mechanistic changeover at $\sigma^* \sim 0.5$ (see fig. G). One has a large negative ρ value, -3.7, indicating $A_{Ac}1$; for the other, ρ is 0.5, for which the mechanism may be $A_{Al}1$, or $A_{Ac}2$.

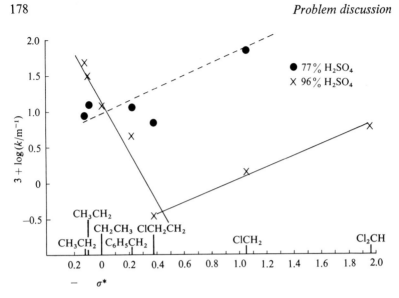

Fig. G.

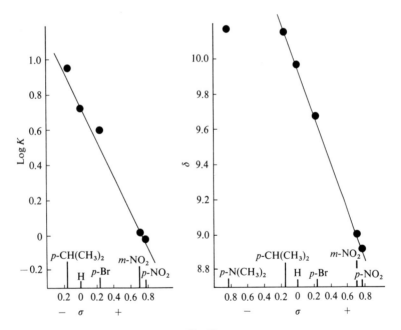

Fig. H.

20. Both log K $\left(K = \dfrac{[\text{open chain form}]}{[\text{ring form}]}\right)$ and δ give good linear plots *vs.* σ (fig. H).

The ρ value for the former is -1.0; electron-withdrawing groups increase the positivity of the carbon atom double-bonded to nitrogen, and thus increase the stability of the ring tautomer. The extent of the chemical shift of the hydroxyl proton is dependent on the strength of its intramolecular hydrogen bonding to N, which registers the availability of the lone pair of this atom, in turn influenced by the electronic character of group X. The point for $N(CH_3)_2$ falls off the line, probably because here the preferred hydrogen bonded form is intermolecular, between the OH group of one molecule and the $N(CH_3)_2$ group of a second.

21.

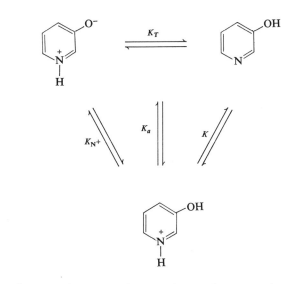

$$\log K_T = \log K_{N^+} - \log K = 0 \qquad \log K_{N^+} = \log K$$

Therefore

$$(\sigma_{N^+} \times 2.1) - 9.99 = (6.0 \times 0.12) - 5.21$$

assuming σ_m for OH to be the same as that for OCH_3. Thus $\sigma_N^+ = 2.6$, a little higher than the values given in table 4.1, illustrating the errors involved in this type of estimation.

22. The data in tables 4.5 and 4.6 enable calculation of the appropriate σ values. For example, σ for 4'-NO_2 in the biphenyl 4-carboxylic acid is $1.23/5 + 1.14/31 = 0.29$, while σ for 6-OCH_3 in the 2-naphthyl systems is $0.20/3.60 - 2.58/17 = -0.09$. The complete calculations afford:

4'-NO_2	0.29	6-NO_2	0.41
3'-NO_2	0.23	7-NO_2	0.36
4'-Br	0.13	6-Cl	0.14
4'-OCH_3	−0.04	7-Cl	0.19
		6-OCH_3	−0.09
		7-OCH_3	0.06

These yield the correlations shown in fig. I.

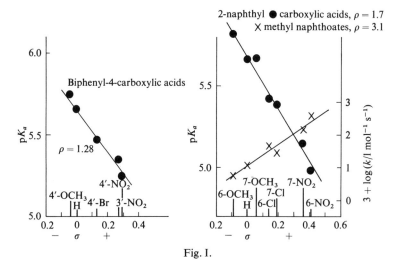

Fig. I.

The excellent agreement shown for the biphenyl results is to an extent fortuitous, because resonance interactions will be smaller than calculated, owing to the twisting of one phenyl unit relative to the other (see § 4.6).

23. From scheme [*4.2*], $K_T = K_1/K_N$, and thus $\log K_T = \log K_1 - \log K_N$ therefore $-1.40 = -pK_1 + 3.44$, $pK_1 = 4.84$. $-4.84 + 5.21 = 6.1 \times \sigma_p$, where 5.21 and 6.1 are the pK_a of pyridine and the ρ value for correlation of pyridine pK_a's respectively. Hence σ_p for $CO_2^- = 0.06$; the resonance withdrawal present in COOH is much reduced on ionisation, while the $-I$ effect will change to $+I$.

24. A reasonable mechanism for the alkaline hydrolysis of 2-methoxy tropones is

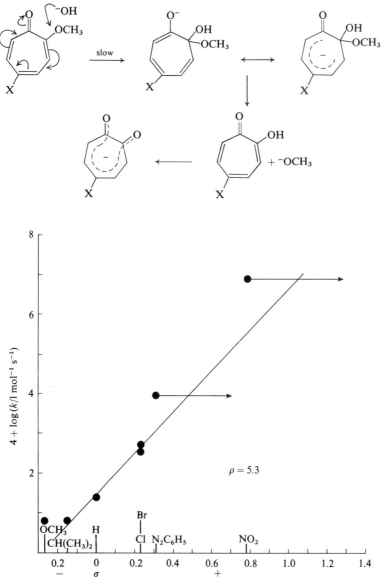

Fig. J.

Since the group X stands '*para*' to both the carbonyl and the methoxy group, a correlation of log k with σ_p is expected. The plot shows a number of interesting features. Firstly, the substituents NO_2 and $N{=}NC_6H_5$ fall off the line in a direction which implies that they are more electron withdrawing than σ_p suggests, which is a reasonable deduction since the transition state for the slow step must involve a ring bearing a partially negative charge. The appropriate value for r in the Yukawa–Tsuno equation (3.14) is thus ~0.5. The point for OCH_3 is also off the line; a better fit is produced using a σ^0_p value for this substituent of 0.14, reflecting the situation which occurs in the ionisation of phenols (§ 2.3).

25. The σ values for the 4-substituents will be equivalent to σ_m; those for the 6- and 8-substituents may be calculated by the usual method. The following values are thus obtained: 4-aza, 0.65; 4-Cl, 0.37; 4-OCH_3, 0.12; 6-CH_3, −0.08; 8-NO_2, 0.52. Figure K shows the excellent correlation produced, the ρ value being 1.78, close to that for the ionisation of benzoic acids, which is 1.67 in 40 % aq. C_2H_5OH. This is commensurate with the deductions of equations (4.6) to (4.9), and suggests that there is

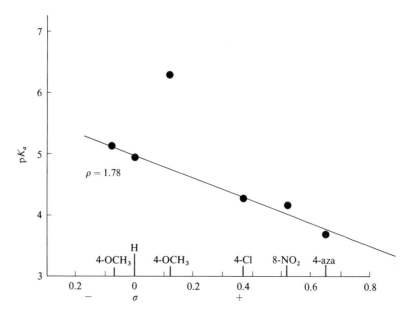

Fig. K.

little interaction, either steric or electronic, between the 1-aza and the 2-COOH group. It also implies that the neutral species is the predominant tautomeric form of the acid, which accords with the infrared data. The one deviant point is for the 4-OCH$_3$ derivative, which also shows no infrared band to indicate an undissociated carboxyl group; here the electron-donating property of OCH$_3$ leads to zwitterion as the predominant tautomer:

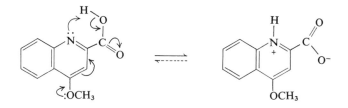

26. A plot of log k *vs.* σ^- at 100 °C produces a good straight line (see fig. L). The ρ value is 5.6, large and positive as expected. A graph of log k *vs.* F gives a very poor correlation, however (see fig. M) so that the results certainly do not obey the isokinetic relationship. Equation (5.12) may be written as

$$\log k_1 = E/2.303R \left(\frac{1}{T_2} - \frac{1}{T_1} \right) + \log k_2$$

enabling the ready calculation of rate constants k_1 at temperature T_1 when the rate constants k_2 at T_2 are known. Application of this equation yields the following results:

Substituents	$9 + \log (k/\text{l mol}^{-1}\,\text{s}^{-1})$, 60 °C	$9 + \log (k/\text{l mol}^{-1}\,\text{s}^{-1})$, 140 °C
H	−1.71	2.91
3-CF$_3$	1.65	5.04
3,5-(CF$_3$)$_2$	4.49	7.35
4-CF$_3$	3.00	5.88
3-SO$_2$CH$_3$	2.88	6.24
4-SO$_2$CH$_3$	5.02	8.08
3-NO$_2$	2.83	6.50
4-NO$_2$	6.81	9.35
3,5-(NO$_2$)$_2$	7.09	9.87
3-CF$_3$, 5-NO$_2$	5.93	8.79
3-SO$_2$CH$_3$, 5-NO$_2$	6.68	9.52

The first graph shows that no significant deviation from the Hammett equation for the temperature range of 80 °C can be observed, despite the non-adherence of the reaction to the isokinetic relationship. This

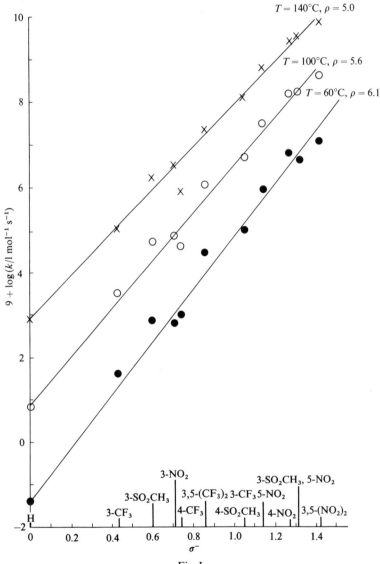

Fig. L.

temperature range is larger than could ever be accurately covered experimentally.

The free energy change $\delta_X \Delta G^{\ddagger}$ for 100 °C for substituent variation H to 3,5-$(NO_2)_2$ is, from equation (5.29), 14×4.18 kJ mol^{-1} arising from a difference in rates of $\sim 10^8$; however no curvature can be detected in the log k, σ plots. Thus if ρ is an authentic measure of transition state structure, the Hammond postulate is inapplicable. The observation of ρ values of 3.6 and 8.5 in the allied reactions with 1-chloro-2-nitrobenzene and chlorobenzene respectively is apparently explained by the Hammond postulate in that the smaller ρ value accords with the faster reaction, which should have a transition state nearer to the reactant molecule; however, the fallacy of this argument is revealed by the fact that 1-chloro-4-nitrobenzene, whose rate is correlated with a ρ of 8.5, reacts at a

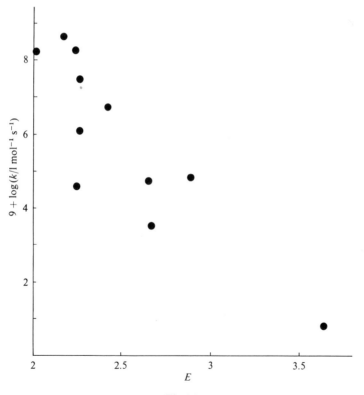

Fig. M.

similar speed to 1-chloro-2-nitrobenzene. A constant proximity or steric effect in the substituted 1-chloro-2-nitrobenzenes altering the transition state structure is the only explanation in accordance with Hammett equation theory.

27. Calculation of ρ (X = OCH$_3$) for reaction:

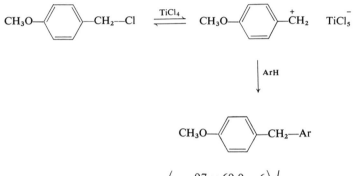

$$\rho = \log \text{(partial rate factor)}/\sigma^+ = \left(\log \frac{97 \times 69.9 \times 6}{100} \right) / -0.31 = -8.4$$

Hence X = OCH$_3$, $\rho = -8.4$; X = CH$_3$, $\rho = -6.6$; X = H, $\rho = -4.3$;
X = NO$_2$, $\rho = -2.3$.

We may write for reaction 1: $(\log k)_1 = \sigma^+ \rho_1 + (\log k_0)_1$

and for reaction 2: $(\log k)_2 = \sigma^+ \rho_2 + (\log k_0)_2$

So that when $(\log k)_1 = (\log k)_2$, $\sigma^+ = \dfrac{(\log k_0)_2 - (\log k_0)_1}{(\rho_1 - \rho_2)}$

Taking reaction 1 as X = NO$_2$, and 2 as X = OCH$_3$:

$$\sigma^+ = -2/6.1 = -0.33$$

i.e. assuming the *Hammett equation to be upheld* for these reactions, benzene substituted with a group of $\sigma^+_{\rho} = -0.33$ would react at equal rates with NO$_2$C$_6$H$_4$$\overset{+}{\text{C}}H_2$ and CH$_3$OC$_6$H$_4$$\overset{+}{\text{C}}H_2$ in the *p*-position. For more reactive aromatics, NO$_2$C$_6$H$_4$$\overset{+}{\text{C}}H_2$ would be a *less* reactive electrophile than CH$_3$OC$_6$H$_4$$\overset{+}{\text{C}}H_2$, so that the selectivity–reactivity relationship breaks down (fig. N).

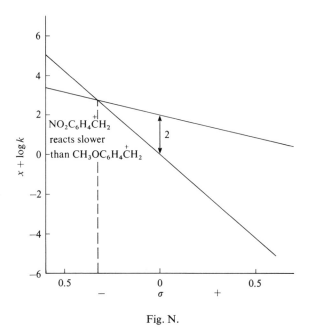

Fig. N.

For the acylation reactions: $X = OCH_3$, $\rho = -9.9$; $X = NO_2$, $\rho = -7.9$. Equal reactivity occurs when $\sigma^+ = -2/2.0 = 1.0$.

The correct deduction might be that ρ is *similar* for all the benzylations and the benzoylations (note the equal ρ values found for reaction of substituted anilines with 1-chloro-2,4-dinitrobenzene and 1,4-dichloro-2-nitrobenzene, § 5.6), but the rates of reaction *differ markedly*, producing the variation in $k_{\text{toluene}}/k_{\text{benzene}}$ due to diffusion control with the faster reacting electrophiles in the competitive reactions (see § 2.7). Absolute reaction rates for electrophile with aromatic would be impossible to find accurately, as their determination requires a knowledge of the concentration of incipient carbonium or acylium ions.

References

Altschuler, L. & Berliner, E. (1966). *J. Am. chem. Soc.* **88**, 5837.
Baker, F. W., Parish, R. C. & Stock, L. M. (1967). *J. Am. chem. Soc.* **89**, 5677.
Bartlett, P. D. (1970). *Q. Rev. chem. Soc.* **24**, 473.
Bender, M. L. & Chen, M. C. (1963). *J. Am. chem. Soc.* **85**, 30.
Bentley, M. D. & Dewar, M. J. S. (1970). *J. Am. chem. Soc.* **92**, 3996.
Berliner, E. & Blommers, E. A. (1960). *J. Am. chem. Soc.* **82**, 6427.
Beverley, G. M. & Hogg, D. R. (1971). *J. chem. Soc.* B, 175.
Biggs, A. I. & Robinson, R. A. (1961). *J. chem. Soc.* 388.
Bishop, D. M. & Craig, D. P. (1963). *Molec. Phys.* **6**, 139.
Bitterwolf, T. E., Linder, R. E. & Ling, A. C. (1970). *J. chem. Soc.* B, 1673.
Blanch, J. H. (1966). *J. chem. Soc.* B, 937.
Bolton, P. D., Fleming, K. A. & Hall, F. M. (1972). *J. Am. chem. Soc.* **94**, 1033.
Bolton, P. D. & Hall, F. M. (1969). *J. chem. Soc.* B, 259.
Bolton, P. D. & Hepler, L. G. (1971). *Q. Rev. chem. Soc.* **25**, 521.
Bosco, M., Forlani, L. & Todesco, P. E. (1970). *J. chem. Soc.* B, 1742.
Bowden, K. & Cook, R. S. (1971). *J. chem. Soc.* B, 1765.
Bowden, K. & Price, M. J. (1971). *J. chem. Soc.* B, 1784.
Bowden, K. & Shaw, M. J. (1971). *J. chem. Soc.* B, 161.
Branch, G. E. K. & Calvin, M. (1941). *The Theory of Organic Chemistry.* New York: Prentice-Hall.
Brown, C. & Hogg, D. R. (1968). *J. chem. Soc.* B, 1262.
Brown, H. C. & Wirkkala, R. A. (1966). *J. Am. chem. Soc.* **88**, 1447.
Brownlee, R. T. C., Hutchinson, R. E. J., Katritzky, A. R., Tidwell, T. T. & Topsom, R. D. (1968). *J. Am. chem. Soc.* **90**, 1757.
Bryson, A. (1960). *J. Am. chem. Soc.* **82**, 4871.
Bunnett, J. F. (1969). In *Survey of Progress in Chemistry*, vol. 5, ed. A. F. Scott. New York: Academic Press.
Butler, A. R. (1970). *J. chem. Soc.* B, 867.
Butler, A. R. & Hendry, J. B. (1970). *J. chem. Soc.* B, 848.
Campbell, A. D., Chooi, S. Y., Deady, L. W. & Shanks, R. A. (1970). *Austr. J. Chem.* **23**, 203.
Charton, M. (1969). *J. Am. chem. Soc.* **91**, 615, 6649.
Ciranni, G. & Clementi, S. (1971). *Tetrahedron Lett.* 3833.
Clementi, S., Linda, P. & Marino, G. (1970). *J. chem. Soc.* B, 1153. *Tetrahedron Lett.* 1389.
Coombes, R. G., Crout, D. H. G., Hoggett, J. G., Moodie, R. B. & Schofield, K. (1970). *J. chem. Soc.* B, 347.
Coombes, R. G., Moodie, R. B. & Schofield, K. (1968). *J. chem. Soc.* B, 800.
Corbett, J. F., Feinstein, A., Gore, P. H., Reed, G. L. & Vignes, E. C. (1969). *J. chem. Soc.* B, 974.
Deady, L. W., Foskey, D. J. & Shanks, R. A. (1971). *J. chem. Soc.* B, 1962.
Dewar, M. J. S. (1969). *The Molecular Orbital Theory of Organic Chemistry.* New York: McGraw-Hill.

Dewar, M. J. S. & Grisdale, P. J. (1962). *J. Am. chem. Soc.* **84**, 3539, 3548.
Donaldson, C. W. & Joullié, M. M. (1968). *J. org. Chem.* **33**, 1504.
Dubois, J-E., Alcais, P. & Rothenberg, F. (1968). *J. org. Chem.* **33**, 439.
Dubois, J-E. & Mouvier, G. (1965). *Tetrahedron Lett.* 1629.
Eaborn, C. (1956). *J. chem. Soc.* 4858.
Eaborn, C. & Fischer, A. (1969). *J. chem. Soc.* B, 152.
Eaborn, C., Eastmond, R. & Walton, D. R. M. (1971). *J. chem. Soc.* B, 127.
Eliel, E. L. (1965). *Angew. chem. internat. Edit.* **4**, 761.
Evans, D. P., Gordon, J. J. & Watson, H. B. (1937). *J. chem. Soc.* 1430.
Exner, O. (1966). *Coll. Czech. chem. Commun.* **31**, 65.
Fang, F. T., Kochi, J. K. & Hammond, G. S. (1958). *J. Am. chem. Soc.* **80**, 563.
Fischer, A., Galloway, W. J. & Vaughan, J. (1964). *J. chem. Soc.* 3591, 3596.
Fischer, A., Hickford, R. S. H., Scott, G. R. & Vaughan, J. (1966). *J. chem. Soc.* B, 466.
Fischer, A., Mitchell, W. J., Ogilvie, G. S., Packer, J., Packer, J. E. & Vaughan, J. (1958). *J. chem. Soc.* 1426.
Fringuelli, F., Marino, G. & Savelli, G. (1969). *Tetrahedron*, **25**, 5815.
Gash, K. B. & Yuen, G. U. (1969). *J. org. Chem.* **34**, 720.
Gleicher, G. J. (1968). *J. org. Chem.* **33**, 332.
Golden, R. & Stock, L. M. (1966). *J. Am. chem. Soc.* **88**, 5928.
Greizerstein, W., Bonelli, R. A. & Brieux, J. A. (1962). *J. Am. chem. Soc.* **84**, 1026.
Hammett, L. P. (1940). *Physical Organic Chemistry*. New York: McGraw-Hill.
Hammond, G. S. (1955). *J. Am. chem. Soc.* **77**, 334.
Hancock, C. K., Meyers, E. A. & Yager, B. J. (1961). *J. Am. chem. Soc.* **83**, 4211.
Harris, J. M., Schadt, F. L., Schleyer, P. von R. & Lancelot, C. J. (1969). *J. Am. chem. Soc.* **91**, 7508.
Hart, H. & Sedor, E. A. (1967). *J. Am. chem. Soc.* **89**, 2342.
Hepler, L. G. (1971). *Can. J. chem.* **49**, 2803.
Hill, E. A., Gross, M. L., Stasiewicz, M. & Manion, M. (1969). *J. Am. chem. Soc.* **91**, 7381.
Hirst, J. & Una, S. J. (1971). *J. chem. Soc.* B, 2221.
Hoggett, J. G., Moodie, R. B., Penton, J. R. & Schofield, K. (1971). *Nitration and Aromatic Reactivity*. London: Cambridge University Press.
Holtz, H. D. & Stock, L. M. (1964). *J. Am. chem. Soc.* **86**, 5188.
Hopkinson, A. C. (1969). *J. chem. Soc.* B, 861.
Humffray, A. A. & Ryan, J. J. (1969). *J. chem. Soc.* B, 1138.
Humffray, A. A., Ryan, J. J., Warren, J. P. & Yung, Y. H. (1965). *Chem. Commun.* 610.
Ingold, C. K. (1953). *Structure and Mechanism in Organic Chemistry*, 1st edition. New York: Cornell University Press.
Ingold, C. K. (1969). *Structure and Mechanism in Organic Chemistry*, 2nd edition. London: Bell.
Ingold, C. K. & Nathan, W. S. (1936). *J. chem. Soc.* 222.
Jackman, L. M. & Kelly, D. P. (1970). *J. chem. Soc.* B, 102.
Jaffé, H. H. (1953). *Chem. Rev.* **53**, 191.
Jaffé, H. H., Freedman, L. D. & Doak, G. O. (1953). *J. Am. chem. Soc.* **75**, 2209.
Jaffé, H. H. & Lloyd Jones, H. (1964). In *Advances in Heterocyclic Chemistry*, vol. 3, ed. A. R. Katritzky. London: Academic Press.
Jencks, W. P. (1964). In *Progress in Physical Organic Chemistry*, vol. 2, ed. A. Streitwieser & R. W. Taft. New York: Wiley.
Jensen, F. R., Bushweller, C. H. & Beck, B. H. (1969). *J. Am. chem. Soc.* **91**, 344.
Johnson, S. L. (1967). In *Advances in Physical Organic Chemistry*, vol. 5, ed. V. Gold. London: Academic Press.
Kershaw, D. N. & Leisten, J. A. (1960). *Proc. chem. Soc.* 84.
Kirkwood, J. G. & Westheimer, F. H. (1938). *J. phys. Chem.* **6**, 506, 513.

Kirsch, J. F., Clewell, W. & Simon, A. (1968). *J. org. Chem.* **33**, 127.
Knowles, J. R., Norman, R. O. C. & Prosser, J. H. (1961). *Proc. chem. Soc.* 341.
Kondo, Y., Matsui, T. & Tokura, N. (1969). *Bull. chem. Soc. Japan*, **42**, 1039.
Kwart, H. & Miller, L. J. (1961). *J. Am. chem. Soc.* **83**, 4552.
Larson, J. W. & Hepler, L. G. (1969). In *Solute–Solvent Interactions*, ed. J. F. Coetzee & C. D. Ritchie. London: Marcel Dekker.
Leffler, J. E. & Grunwald, E. (1963). *Rates and Equilibria of Organic Reactions.* New York: Wiley.
Love, P., Cohen, R. B. & Taft, R. W. (1968). *J. Am. chem. Soc.* **90**, 2455.
McDaniel, D. H. & Brown, H. C. (1958). *J. org. Chem.* **23**, 420.
McDonagh, A. F. & Smith, H. E. (1968). *J. org. Chem.* **33**, 1.
Mouvier, G. & Dubois, J-E. (1968). *Bull. Soc. chim. France*, 1441.
Norman, R. O. C., Radda, G. K., Brimacombe, D. A., Ralph, P. D. & Smith, E. M. (1961). *J. chem. Soc.* 3247.
Norman, R. O. C. & Taylor, R. (1965). *Electrophilic Substitution in Benzenoid Compounds.* London: Elsevier.
Olah, G. A. (1971). *Accs. of chem. Res.* **4**, 240.
Ostrogovich, G., Csunderlik, C. & Bacaloglu, R. (1971). *J. chem. Soc.* B, 18.
Petersen, R. C. (1964). *J. org. Chem.* **29**, 3133.
Pietra, F. (1969). *Q. Rev. chem. Soc.* **23**, 504.
Poutsma, M. L. (1965). *J. Am. chem. Soc.* **87**, 4285.
Price, C. C., Mertz, E. C. & Wilson, J. (1954). *J. Am. chem. Soc.* **76**, 5131.
Rao, C. N. R. (1963). *Chemical Applications of Infrared Spectroscopy.* New York: Academic Press.
Reid, E. E. (1900). *J. Am. chem. Soc.* **24**, 397.
Renfrow, W. B. & Hauser, C. R. (1937). *J. Am. chem. Soc.* **59**, 2308.
Ridd, J. H. (1971). *Accs. of chem. Res.* **4**, 248.
Ritchie, C. D. & Lewis, E. S. (1962). *J. Am. chem. Soc.* **84**, 591.
Ritchie, C. D. & Sager, W.F. (1964).In *Progress in Physical Organic Chemistry*, vol. 2, ed. S. G. Cohen, A. Streitwieser & R. W. Taft. New York: Wiley.
Roberts, J. D. & Moreland, W. T. (1953). *J. Am. chem. Soc.* **75**, 2167.
Rogne, O. (1970). *J. chem. Soc.* B, 727.
Ruskie, H. E. & Kaplan, L. A. (1965). *J. org. Chem.* **30**, 319.
Russell, G. A. (1958). *J. org. Chem.* **23**, 1407.
Schaefer, J. P. & Miraglia, T. J. (1964). *J. Am. chem. Soc.* **86**, 64.
Sheppard, W. A. (1963). *J. Am. chem. Soc.* **85**, 1314.
Skell, P. S. & Cholod, M. S. (1969). *J. Am. chem. Soc.* **91**, 7131.
Šmejkal, J., Jonáš, J. & Farkaš, J. (1964). *Coll. Czech. chem. Commun.* **29**, 2950.
Smith, G. G. & Kelly, F. W. (1971). In *Progress in Physical Organic Chemistry*, vol. 8, ed. A. Streitwieser & R. W. Taft. New York: Wiley.
Stewart, R. & Walker, L. G. (1957). *Can. J. Chem.* **35**, 1561.
Stock, L. M. & Brown, H. C. (1963). In *Advances in Physical Organic Chemistry*, vol. 1, ed. V. Gold. London: Academic Press.
Streitwieser, A. (1962). *Molecular Orbital Theory for Organic Chemists.* New York: Wiley.
Streitwieser, A., Mowery, P. C., Jesaitis, R. G. & Lewis, A. (1970). *J. Am. chem. Soc.* **92**, 6529.
Swain, C. G. & Lupton, E. C. (1968). *J. Am. chem. Soc.* **90**, 4328.
Taft, R. W. (1956). In *Steric Effects in Organic Chemistry*, ed. M. S. Newman. New York: Wiley.
Taft, R. W. (1960). *J. phys. Chem.* **64**, 1805.
Taft, R. W. & Lewis, I. C. (1958). *J. Am. chem. Soc.* **80**, 2436.

Taft, R. W., Price, E., Fox, I. R., Lewis, I. C., Andersen, K. K. & Davis, G. T. (1963). *J. Amer. chem. Soc.* **85**, 709, 3146.

Tommila, E. (1941). *Ann. Acad. Sci. Fennicae*, **57**, 3.

Tommila, E. & Hinshelwood, C. N. (1938). *J. chem. Soc.* 1801.

Totherow, W. D. & Gleicher, G. J. (1969). *J. Am. chem. Soc.* **91**, 7150.

Ueno, Y. & Imoto, E. (1967). *J. chem. Soc. Japan*, **88**, 1210.

Van Bekkum, H., Verkade, P. E. & Wepster, B. M. (1959). *Rec. Trav. chim. Pays-Bas*, **78**, 815.

Wells, P. R. (1963). *Chem. rev.* **63**, 171.

Wells, P. R., Ehrenson, S. & Taft, R. W. (1968). In *Progress in Physical Organic Chemistry*, vol. 6, ed. A. Streitwieser & R. W. Taft. New York: Wiley.

Wheland, G. W., Brownell, R. M. & Mayo, E. C. (1948). *J. Am. chem. Soc.* **70**, 2492.

Wilcox, C. F. & Leung, C. (1968). *J. Am. chem. Soc.* **90**, 336.

Williamson, K. L., Jacobus, N. C. & Soucy, K. T. (1964). *J. Am. chem. Soc.* **86**, 4021.

Woodward, R. B. & Hoffmann, R. (1969). *Angew. chem. Internat. Edit.* **8**, 781.

Yates, K. & McClelland, R. A. (1967). *J. Am. chem. Soc.* **89**, 2686.

Yukawa, Y. & Tsuno, Y. (1959). *Bull. chem. Soc. Japan*, **32**, 971.

Yukawa, Y., Tsuno, Y. & Sawada, M. (1966). *Bull. chem. Soc. Japan*, **39**, 2274.

Index